SpringerBriefs in Statistics

More information about this series at http://www.springer.com/series/8921

Warren Beatty

Decision Support Using Nonparametric Statistics

Springer

Warren Beatty
University of South Alabama
Mobile, AL, USA

ISSN 2191-544X ISSN 2191-5458 (electronic)
SpringerBriefs in Statistics
ISBN 978-3-319-68263-1 ISBN 978-3-319-68264-8 (eBook)
https://doi.org/10.1007/978-3-319-68264-8

Library of Congress Control Number: 2017955828

Printed on acid-free paper

This Springer imprint is published by Springer Nature
The registered company is Springer International Publishing AG
The registered company address is: Gewerbestrasse 11, 6330 Cham, Switzerland

Decision Support Using Nonparametric Statistics

This E-Book utilizes OpenOffice, a FREE, downloadable spreadsheet program

ALL text examples and end-of-chapter exercises data and solutions are in downloadable spreadsheets that facilitates examination and understanding.

ALL the figures in this text are OpenOffice spreadsheets that can be downloaded. So don't worry if you can't read the figures. You can download the chapter examples spreadsheets—the figures, open them with OpenOffice, and see how the nonparametric statistic was accomplished.

The text was designed to be used WITH a computer. For Chaps. 7–10 all the text figures are downloadable spreadsheets that can be opened with OpenOffice. By the way, all exercise solutions can be downloaded as well.

Decision Support Using Nonparametric Statistics

Warren Beatty

Dedication

To Al and Carl: Teachers, Mentors, Friends

Decision Support Using Nonparametric Statistics

Using a FREE Spreadsheet

Statistical thinking will one day be as necessary for efficient citizenship as the ability to read and write.
Herbert George (H.G.) Wells (1866–1946) Author

---

or, to paraphrase Wells,
THIS IS GOOD STUFF!

Preface to Decision Support Using Nonparametric Statistics

I always find that statistics are hard to swallow and impossible to digest. The only one I can ever remember is that if all the people who go to sleep in church were laid end to end they would be a lot more comfortable!
Mrs. Robert A. Taft (Martha Wheaton Bowers), daughter-in-law of W.H. Taft, 27th US President

Like other occult techniques of divination, the statistical method has a private jargon deliberately contrived to obscure its methods from non-practitioners.—G.O. Ashley

Purpose

This book is designed for managers and students who want to know about nonparametric statistics. To be a knowledgeable consumer of the world's use of nonparametric statistics, one must understand how nonparametric statistics can support critical decisions. One needs very little knowledge of mathematics in order to understand the material in this book. Computer packages, such as the statistics capabilities in OpenOffice, have almost eliminated the need for computational formulae. There is a Windows version and an Apple Mac version that can be downloaded.

Windows version: http://www.openoffice.org/download/
Apple version: https://www.openoffice.org/porting/mac/

What formulae are used can easily be copied from downloaded spreadsheets.

This book is written so as to free the manager to examine WHY the data analysis suggests a decision without getting "bogged down" with how the analysis is accomplished.

It also frees the "statistics instructor" from having to tell students HOW to generate statistical output to be discussed. It allows the instructor to concentrate on what he or she does best—convey nonparametric statistical concepts by using illustrations students can generate for themselves.

What is Nonparametric Statistics and how does it differ from "regular" (parametric) Statistics?

Nonparametric statistics make FEWER assumptions regarding the underlying distributions of populations than do "regular" parametric statistics. For example, most regular statistics procedures require assumptions about the population from which the samples are drawn be normally distributed, something most nonparametric statistics procedures do not make. Further, a nonparametric procedure MUST be used if the population (from which the sample is drawn) is skewed or otherwise not normally distributed, and/or sample data were measured using the nominal or ordinal measurement scales.

Nonparametric statistics is often called **DISTRIBUTION-FREE STATISTICS** because no (population) distribution assumptions are made.

There is a quote about statistics attributed to Benjamin Disraeli: "Lies, damned lies, and statistics." Disraeli was referring the use of statistics to strengthen weak arguments. Mark Twain expanded upon Disraeli's remark: "There are three kinds of lies: lies, damned lies, and statistics." With that introduction to statistics, you had better understand what is going on in order to protect yourself, just like the last instruction the referee gives to boxers: "Protect yourself at all times."

As Twain DIDN'T say: "liars, outliers, and out-and-out liars"! (Nor did Disraeli...).—Robert Dawson

This book will help you:

- Understand the nonparametric statistics presented in your world in particular and in the world in general
- Communicate to others how nonparametric statistical information supports your decision
- Understand the relationship between nonparametric statistics and decision support
- Apply this new knowledge of nonparametric statistics and use these helpful new tools **to improve your decision-making**

Intended Audience

What is the difference between an extroverted and an introverted statistician? The extroverted statistician looks at **your** shoes while talking to you.—Jim Box

Let's be honest: Statistics can be boring, can intimidate, confuse, and mislead people into believing things that are not true. Have you ever heard the phrase *caveat emptor*—let the buyer beware? This book, by arming you with the decision-making knowledge provided by nonparametric statistics, will let you (the buyer) beware (therefore avoid) using inappropriate data analysis tools to make decision. It will also help by removing some of the confusion of statistics, and by helping you recognize misleading information. It will keep you from making a disastrous decision, as well as keep you from "buying a pig in a poke."

Perhaps you are not reading this book by choice, but by necessity. If you are a student who is forced to further your statistics knowledge, this book is written with you in mind. And you will see that a knowledge of nonparametric statistics is quite easy to obtain, useful to have, and can assist you when you have to make everyday decisions.

This book is for you if you are wary of statistics, mathematics, or of computers and spreadsheets. This book will help anyone who wants to learn how to apply and interpret the OpenOffice statistical tools in order to generate decision-making information, and how to interpret the information generated. This book illustrates how to conduct nonparametric statistical tests and how to interpret the results.

Anyone majoring in business, psychology, nursing, education, sociology, engineering, or computer science who wants to learn how to use OpenOffice to solve and interpret nonparametric statistics problems can benefit. This book is appropriate for use in ANY introductory nonparametric statistics course, whether undergraduate or graduate.

While this book is aimed primarily at managers who want to use a spreadsheet to generate information in support of decision-making, it can be useful to students as well. Managers, after all, are students themselves, and the learning process never ends!

Scope—Topic Coverage

It is simply not possible to acquire a complete understanding of nonparametric statistics via a single introductory course. Therefore, what this book does is introduce nonparametric statistics topics that are most likely to be seen and used from a decision support perspective:

- Introduction to the Process of Decision-Making
- Some Basic Statistics Concepts
- Introduction to Variables and Variable Characteristics
- Introduction to Probability and Binomial Probability
- Introduction to Hypothesis Testing and its Logic
- One Population Hypothesis Testing
- Two Population Hypothesis Testing
- Three (or more) Population Hypothesis Testing
- Association

Don't like the order of presentation? Chapters 5 and 6 introduce some fundamental statistics concepts. All other chapters are "stand alone" so the topics they cover can be addressed in the order that best suits the instructor's needs.

Why You Need This Book

It is proven that the celebration of birthdays is healthy. Statistics show that those people who celebrate the most birthdays become the oldest.—S. den Hartog

Meaningless statistics are up 6.5% from last year.

This book is a palatable introduction to BASIC nonparametric statistics. It focuses upon applications using a spreadsheet with statistics capabilities to generate information, and upon interpretation of results in support of decision-making, rather than upon formulae. It introduces the most used nonparametric statistical tools or procedures that can be useful with any discipline. It prepares you to be a consumer of statistics in the world.

ALL spreadsheets, including chapter illustrations, chapter exercises, and chapter exercise answer, can be found in the "Chapter Examples, exercise data, and exercise solutions spreadsheets downloads" section of the Table of Contents. They can easily be downloaded so you can see how the chapter illustrations were actually calculated, work with consistent data, and check/verify your exercise answers!

OpenOffice is a computer spreadsheet program that is a Microsoft Excel look-alike and work-alike, and can be downloaded for FREE, **at no cost to you.** EXCEL costs quite a bit of money. Therefore, buying this book can save you a lot of money, both initially by not having to buy EXCEL and by making you a better decision maker. Additionally, all the functions and tools in this book are also applicable to Excel users. Do not worry if you have already bought Excel: everything in this book is applicable.

This is the first book to use the capability of OpenOffice to teach nonparametric statistics. You will learn how to use OpenOffice to perform nonparametric statistical procedures without getting swamped or overpowered by nonparametric statistical theory or formulae. The best part is that the OpenOffice spreadsheets are used to generate nonparametric statistics examples, applications, and illustrations throughout the entire book. In addition, this book, in the OpenOffice Tutorial section, includes instructions for downloading, installing, and using your own copy of OpenOffice.

You get what you pay for—period. However OpenOffice may be an exception to that rule. Excel is the most popular and widely sold spreadsheet and statistics software in the world today. Learning to use OpenOffice can get you ready if you ever are asked if you know how to use Excel (like at a job interview), or if you are asked if you know how to use and/or interpret nonparametric statistics.

There are literally thousands of nonparametric statistics resources available on the Internet, but those resources are inconsistent and make use of many available statistics programs. There is a lot to be said for a consistent presentation that uses only one statistics program.

There are, throughout this book, humorous asides that I hope readers will find as amusing as I do. "If we couldn't laugh at ourselves, that would be the end of everything."—Niels Bohr, Danish physicist

If you are so inclined, I would encourage you to check out

Rob J. Hyndman's Statistical Jokes

Gary Ramseyer's World of Statistical Humor!

Joachim Verhagen's Science Humor—For some reason, I like 9.5, The Nerd Test

Statistics Jokes

Don Zimmerman's Statistical Devil's Dictionary

WorkJoke—funny statistician jokes. Well, at least ***I*** thought they were funny!

Mobile, AL, USA Warren Beatty

Acknowledgement

I wish to acknowledge the OpenOffice team for a superior spreadsheet.

Contents

1
The Decision Making Process

How Statistics Supports the Decision Making Process

Decision Making Is Simply Unavoidable

A psychiatrist asks his patient if he has trouble making decisions? The patient responds, "Well, doctor, yes and no."

Decision making can be regarded as a process which results in the selection of a course of action from among several alternative courses of action. Every decision making process produces a final choice, or decision.

Most decisions are made unconsciously. Most people decide without thinking about the decision-making process. Remember, even doing nothing can be considered making a decision.

Consider this from *Entrepreneur* magazine: "Putting a process around decision making can help you guard against your weaknesses and be more effective. Used consistently, a good process can facilitate quick decisions at the right level, train employees on their decision-making skills and produce better results."

From Reference.com comes, "Statistical analysis allows businesses to deal with the uncertainties of the business. It allows managers to make sound judgments, knowing their decisions are based on data and not on assumptions. Statistics helps businesses to plan better and make predictions about the road ahead."

From the two quotes above, it is clear that there is a need for a process to follow, and statistical analysis can facilitate the decision-making process. There are many good decision-making processes, varying from three to nine steps. They all have two things in common: the all require analysis of data and the communication of the results. One process is presented below:

W. Beatty, *Decision Support Using Nonparametric Statistics*, SpringerBriefs in Statistics,
https://doi.org/10.1007/978-3-319-68264-8_1

1. Define the question(s) to be answered: questions should be measurable and clear
2. Determine what and how to measure: thinking about how you measure your data is just as important as what, especially before data collection begins, because your measuring process either backs up or discredits your analysis later on (if there was ever a case for nonparametric statistics, this is it!)
3. Collect the Data: with your question clearly defined and you know what and how to measure, collect your data
4. Analyze the Data: data analysis tools and software are extremely helpful here
5. Interpret the Results of the Analysis: determine what the analysis suggests
6. Make the Decision: implement what the analysis suggests, "pull the trigger"
7. Communicate What You Decided: tell your boss and subordinates WHY you decided as you did

If statistics is so great for supporting decision-making, how specifically does it enhance the decision-making process? To attempt to answer that question, consider what is required of statistics. From John Dillard at Big Sky we get these requirements:

1. You need to know if it is the right data for answering your question;
2. You need to draw accurate conclusions from that data; and
3. You need data that informs your decision making process

As this book illustrates, nonparametric statistics, using its various tools, will accurately analyze any amount and kind of data collected, then validly interpret the results and **suggest**, via a hypothesis test, what decision to make. But it never makes a decision for you.

One of the requirements of good decision making is communication. For the decision itself, wouldn't it be nice to have some substantiation for the decision you must make? Nonparametric statistics, via printable spreadsheets, can provide that substantiation.

Nowhere in this book will you find ***what*** decision to make. All nonparametric statistics can ever do is provide information to support, assist, and improve your decisions after assessing the risks associated with the choice of expected or potential outcomes. The closest to making a decision will come when a particular technique is introduced and its use must be explained.

2
Four Facts About ALL Statistics

There are four very important facts about nonparametric statistics, and all statistics in general, that you should know:

1. Statistics in this book have ABSOLUTELY nothing to do with advanced mathematics. It is unfortunate that statistics has been lumped with advanced mathematics. Some (very) basic knowledge of math is required, but that's all. Until the mid-1970s, complex looking formulas filled the pages of statistics books. All of those formulas were nothing more than computational short cuts, conveniences for people who had to compute values with pencil and paper. Advanced math provided those conveniences. ANY shorter and/or easier formulae were welcome. The computer changed that, freeing people from tedious, error-prone calculations. Statistics books began to change, focusing more on the use of computers to calculate statistics to support the decision-making process, and less on formulae and short-cuts.

2. Statistics NEVER MADE A DECISION! All statistics (of any kind) can ever do is lend support and analysis to the decision that you, the decision-maker, have to make. Make good decisions: you get promoted. Make bad decisions: you get fired. And if anyone asks, you can support your decision by showing the analysis you have done.

3. It is INCUMBENT UPON YOU, the decision maker, to be certain that the correct statistical analysis procedure is used. There never has been, nor will there ever be, a procedure to select the "correct" statistical procedure. The toolbox analogy works well here: Statistics is a toolbox full of tools. It is up to you to select the correct tool and to use it correctly.

4. Because ALL statistics are estimates, they are not EXACTLY what you wish to know. But that fact should not bother you because statistics can still yield good information. Statistics are ESTIMATES of what is really going on in the population. And if you follow what is presented here, the estimates you generate will be very close to the unknown population value that you wish you could know.

W. Beatty, *Decision Support Using Nonparametric Statistics*,
SpringerBriefs in Statistics,
https://doi.org/10.1007/978-3-319-68264-8_2

3
Introduction

Decision Support Information
Information is a source of learning. But unless it is organized, processed, and available to the right people in a format for decision making, it is a burden, not a benefit.
William Pollard (1911–1989) Physicist

This book is about generating information using nonparametric statistical analysis to support decision-making, about using nonparametric statistics to organize and process raw data in order to make available information in a usable and understandable format to support decision making.

WHAT IS STATISTICS?

John Tukey, a famous statistician, once said, "Statistics is a science, not a branch of mathematics, but uses mathematical models as essential tools."

Formally, it is the science of collecting and analyzing data for the purpose of making inferences about an entire population from a representative sample. It is a discipline concerned with the analysis of data and decision making based upon that analysis.

From the University of Melbourne (Australia), we get this definition: Statistics is a discipline which is concerned with:

- designing experiments and other data collection,
- summarizing information to aid understanding,
- drawing conclusions from data, and
- estimating the present or predicting the future. In making predictions, Statistics uses the companion subject of Probability, which models chance mathematically and enables calculations of chance in complicated cases.

W. Beatty, *Decision Support Using Nonparametric Statistics*,
SpringerBriefs in Statistics,
https://doi.org/10.1007/978-3-319-68264-8_3

A famous quote about statistics is attributed to English Prime Minister Benjamin Disraeli, who in 1871 declared, "There are three kinds of lies: lies, damned lies, and statistics." He was referring to the propensity of some politicians for distorting information by carefully selecting data. Politicians may lie, but statistics never does. Statistics works every time ***IF*** the assumptions underlying the statistical theorems are followed. Failure is usually caused by the sample of data collected not being sufficiently random or biased in some other manner.

TWO TYPES OF STATISTICAL PROCEDURES

There are two types of statistical procedures: Parametric (or "regular") statistics procedures, and nonparametric statistical procedures. This book is primarily about nonparametric statistics procedures, about using **nonparametric statistical procedure analysis** to help you make a decision.

The nonparametric statistics procedures introduced here will work for **ANY** discipline in which decisions must be made—which pretty much covers any subject that might come to mind where historical data is available.

> Statistical Analysis: Mysterious, sometimes bizarre, manipulations performed upon the collected data of an experiment in order to obscure the fact that the results have no generalizable meaning for humanity. Commonly, computers are used, lending an additional aura of unreality to the proceedings.

Definition of NonParametric Statistics

What is this "body of knowledge" called Nonparametric Statistics? It is the science that addresses the collection, classification, analysis, and interpretation of numerical and non-numerical facts or data in order to **support decision making.** Nonparametric Statistics identifies the values, uncertainties, and other issues for a given decision situation. It provides information in order to help you make the best possible decision. It formally orders the analysis and decision making processes so you can make objective decisions. Nonparametric statistics relies upon historical data (that you gather) and analyzes that data to provide information. Nonparametric statistics methods make, and are bound by far fewer assumptions about the population(s) from which a sample(s) is taken, and an inference(s) made, than "regular" statistics.

> Remember, if you torture historical data enough, it WILL confess.

Why Use Nonparametric Statistics Procedures?

Nonparametric Statistics can save money and/or time, and, at the same time, can "keep you out of trouble." Nonparametric statistics procedures can require much smaller samples than parametric (or "regular") statistics, and can much more easily and quickly be performed. Nonparametric statistics can keep you out of trouble by providing reliable decision support information when parametric (or "regular") statistics assumptions are not met.

The major advantages of nonparametric statistics are:

- they make fewer assumptions about the nature of the population(s) distribution(s) from which sample(s) are drawn
- they can therefore be applied to a large number of situations and can be used with more types of data
- they require smaller sample sizes
- they are generally not seriously affected by extreme values in data (outliers)
- they can be more easily understood
- they can be used to analyze ANY kind of data regardless of what scale (the next chapter) upon which the data were measured

> If I had only one day left to live, I would live it in my statistics class: it would seem so much longer.—Urban Myth, or is it a myth?

Everyday Decision Situations

You may not realize that you use nonparametric statistics every day to support decisions you make.

- Do consumer satisfaction ratings have any meaning that can be used?
- Will one brand of tire perform better than another?
- Do you go to the effort to look prettier (or more handsome) in order to get more dance invitations?
- Is the bingo game you are playing random (and honest) or not?

All this book does is provide terminology and structure to the nonparametric statistical analysis processes that you are already using in the course of a normal life.

DECISION SITUATION EXAMPLES

In this first decision situation, assume that you are the manager of a delivery company. Tire mileage is of primary interest to you, with superior mileage defined as "success." If a sales person for one particular brand of tire says hers will deliver superior mileage, how do you determine whether she is telling the truth or not?

In the second decision situation, how can you tell if two (or more) different brands of tires provide similar mileage?
In the third decision situation, you (for whatever reason) are trying to see if NASCAR race car numbers has anything to do with qualifying speed.
In the fourth decision situation, you are trying to determine if the satisfaction ratings of two (or more) products are the same or not.
In the fifth decision situation, you are trying to determine if salesperson competence (or effort) has anything to do with customer satisfaction.
In the sixth decision situation, you, as a product manager, are trying to decide whether to continue a specific product based upon consumer ratings.

These six decision examples illustrate what nonparametric statistics CAN do, analyze historical data and provide you, the manager or decision maker, with information to support the decision you make.

Note the first two situations **could** be analyzed using "regular" or parametric statistics because they involve the analysis of ratio data (introduced in the next chapter). However there is nothing incorrect about using nonparametric statistics to analyze the data.

But the third through sixth decision examples involve the analysis of nominal or ordinal data, therefore **require** the use of nonparametric statistics. Use of parametric statistics to analyze this data will yield invalid results and cause bad decisions.

A married couple was celebrating their 60th wedding anniversary. At the party everybody wanted to know how they managed to stay married so long in this day and age. The husband responded "When we were first married we came to an agreement. I would make all the major decisions and my wife would make all the minor decisions. And in 60 years of marriage we have never needed to make a major decision."

STATISTICAL ASSUMPTIONS

All of the nonparametric statistical procedures presented make two assumptions. These assumptions are not about data itself, but about HOW the data are collected. If these two assumptions are not met (or closely met), ANY statistical analysis will prove meaningless, and all your work will be a waste of time and effort. Fortunately, these assumptions are easily met.

First, there is the assumption of **INDEPENDENCE**, which simply means that an outcome of one event (the selection of one datum item from a population) has no effect upon subsequent events. All of these nonparametric statistics procedures work only if this assumption is met. (Note: this is NOT the same concept as Independent and Dependent samples introduced in Chap. 8)

Second, there is the assumption of **RANDOMNESS**. The term random means "equal opportunity of selection." Therefore when selecting from a group, if all group members have an equal chance of being selected, then the randomness assumption is met. There is no "bias," or methodology of group member identification for selection.

Note that nonparametric statistics procedures make no assumption of normality of the population(s) from which a sample(s) is taken. But parametric statistics does.

PROOF versus INFERENCE

You never PROVE anything with statistics. You make an **INFERENCE** (an educated guess) about the entire population (that you defined) based upon your analysis of the sample using statistics. The inference you make will be the basis for your decision!

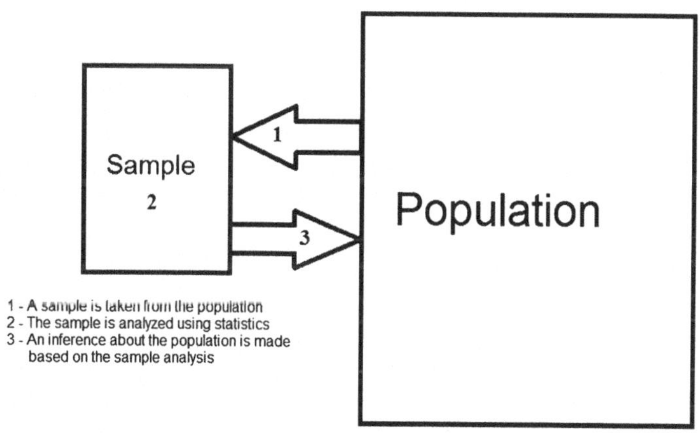

Figure 3.1: Making an inference from a population

What the nonparametric **statistical analysis** (step 2 in Fig. 3.1 above) does is calculate, using the sample, what is known as a "**test statistic**." The test statistic value, what is compared to table values, is an estimate of what is actually the condition (something going on or not) in the entire population. The test statistic value is used in hypothesis testing.

HYPOTHESIS TESTING

The concept of hypothesis testing fits well with proof versus inference. Hypothesis testing is based entirely upon the test statistic value calculated from the sample, so there is ALWAYS a possibility that the information generated by hypothesis testing is wrong. But hypothesis testing is a formal, objective, documentable way to minimize the possibility of an error being made.

The process of Hypothesis Testing is used to generate information to help make a decision about a claim or opinion by addressing the question: "Can the difference between what is observed and what is claimed or opined be attributed only to chance?" H_0 (what is named the "null hypothesis") says that ANY difference between what is claimed or opined and what is observed (in the sample) can be attributed to the fact that a sample is never exactly the same as the population from which it was drawn. The name for this difference is "sampling error." An "error," the difference between the population and the sample drawn from it, is made every time a sample is drawn. There is no way to avoid the situation. But the "error" can be minimized by meeting the two assumptions mentioned above. This situation is true regardless of the number of populations being examined.

The initial claim and/or opinion, H_0, usually states that there is nothing going on (that you don't know about), that any differences between what is claimed or opined and what is actually observed (from sample data) is due only to chance. H_0 is formed this way because you, the decision maker, want the sample evidence (data) to convince you to reject H_0, the opinion that nothing is going on.

This is a very important point. By forming H_0 this way, you are saying "OK, sample, convince me to reject my initial claim or opinion."

H_0 is NEVER proven to be true or correct. You either reject it or not. See this source for more information.

Suppose you are of the opinion that men and women have similar thoughts regarding some issue of importance. Your opinion must be objectively stated:

H_0: men and women have similar thoughts regarding the issue (and any differences between men and women observed in the sample data are due to chance).

Similarly, the alternative to your opinion must be objectively stated: H_1 (named the alternative hypothesis): men and women do not have similar thoughts regarding the issue (and any differences between men and women observed in the sample data are NOT due to chance).

There can never be any overlap or ambiguity regarding expressing H_0 and H_1, what is called "mutually exclusive." Also, every possible outcome must be expressed when the two hypotheses are presented, what is called "collectively exhaustive."

H_0 ALWAYS has an equality (=) in it:
H_0: some test statistic = 0 (or some value based in the initial claim or opinion)
H_1 must state all other value possibilities that can occur: H_1: some test statistic $\neq$ (not equal to) 0

Notice in the above formulation, there is absolutely no overlap or ambiguity, and that ALL possible test statistic values are addressed.

NOTE: all hypothesis tests presented in this book are two-tailed in nature, since that is what is most often encountered with nonparametric statistics.

Rejecting H_0 does not mean that H_1 is proven to be true, only that sample evidence convinces you to abandon or reject your initial claim or opinion.

REMEMBER, nothing is ever proven with statistics! All statistical analysis ever does is generate information to assist the decision maker. Look at the situation this way: rejecting (or not) H_0 is one more piece of information to support the decision that must be made.

SIGNIFICANCE LEVEL

Hypothesis testing permits an expression of how confident you desire to be regarding being correct about rejecting or not rejecting H_0. This is called the "significance level." For example, say you, the decision maker, want to be 95% confident that you are correct regarding the rejection of H_0. The analysis permits you to be 95% confident that the correct decision is being made. But there are two very important caveats of which you must be aware:

1. the significance level, your level of confidence that the correct decision is being made, must be expressed BEFORE any analysis is performed, and cannot be altered if OBJECTIVE decisions are to be made. For

example, if you want to be 95% confident that the correct decision regarding H_0 is being made, then what is named "α", is = 0.05, or 5%, the maximum chance of making an incorrect decision regarding H_0. α is simply (1 – level of confidence). Specify α **BEFORE** starting any analysis, then stick with it!

2. α can NEVER be zero. The lower the value of α, the more confident you are that the correct decision regarding H_0 is being made. But remember, nothing is ever proven with statistics, and that α can never be 0. The significance level can be ANY value you choose, but can never be 100%. And α can never be zero!

Regarding claims; claims (usually by salespeople) form initial opinions. The salesperson may be lying, so analysis of his/her claim is warranted. Similarly, your experience forms your opinion, your claim about what may be going on. Your experience may be misleading, so analysis is warranted.

How can you tell if a salesperson is lying? His/her lips are moving!

FIVE STEPS OF HYPOTHESIS TESTING

There are five steps to ALL hypothesis testing, and they MUST be performed in this order if you are to be OBJECTIVE.

Step 1:	Define and state both H_0 and H_1
Step 2:	Define α (usually 0.05)
Step 3:	Take your sample
Step 4:	Compare your calculated test statistic value
Step 5:	Make your decision

- Step 1: Define and state both H_0 and H_1
- Step 2: Define α, (Remember, α = confidence you're willing to give up) which also defines your "critical" test statistic value and confidence that you are making the correct decision regarding H_0
- Step 3: Take your sample (make an observation), then calculate your test statistic value, usually called Z_{calc} (or some other name—the name does not matter)
- Step 4: Compare your calculated test statistic value that was calculated in Step 3 to the "critical" test statistic value (determined by α that you specified, or the "critical" value found in an appendix table) or compare the p-value of the test statistic to α

- Step 5: Make your decision to either reject or fail to reject H_0. Remember, H_1 in never accepted.

If you complete all five steps in this order, you will be objective and (most times) make good managerial decisions.

Here are some facts about hypothesis testing with which you should be familiar:

- The level of significance for hypothesis testing can be any value from 50% (a coin toss) to 99.99%, but it can never be 100% because an inference is being made. The higher the significance level, the more certain that you are making the correct decision regarding H_0. Similarly, α values can range from 0.5 to 0.0001. Why not, then, always set the level of significance at 99.99% and $\alpha = 0.0001$? There is a reason. Sample evidence must be "very convincing" in order to reject H_0, the initial claim or opinion, if $\alpha = 0.0001$. Unless the sample is very convincing, H_0 will never be rejected, and you will never reject your initial opinion. So, what level of significance should be expressed? That is entirely up to you, the decision maker. The more important the decision, the higher the level of significance. For example, if Daddy owns the company and you will not get fired regardless of the decision made, set the level of significance at 50% and toss a coin. You have a 50% chance of making the correct decision. If, however, your job and continued family support depend upon the correct decision being made, you may want to increase the level of significance.

- What is the "correct" level of significance? Statistics cannot help here. The "correct" level depends upon how confident you, the decision maker, want to be regarding the decision to be made. By convention, and only by convention, the levels of 90% (α=0.1), 95% (α=0.05), and 99% (α=0.01) are most used. There is absolutely nothing sacred about these levels. For most people, a level of significance of 95% is "acceptable." That means that the maximum probability of making an in correct decision is 5% (100% − 95%). Is that probability of making an incorrect decision low enough? Your call!

- What is this thing called a "p-Value?" It is simply the probability of what is observed (the sample) occurred if H_0 cannot be rejected, and the probability of the test statistic value if H_0 cannot be rejected. If the p-value is greater than α, H_0 cannot be rejected. If the p-value is greater than α, then you, the decision maker, conclude that what you observed (the sample) has a greater than α% probability of occurring due to chance. If, however, the p-value is less than α, then H_0 CAN be rejected.

- The term "significant" is often used if H_0 can be rejected. That term simply means that H_0 has been rejected. It has absolutely nothing to do with sample analysis or interpretation.

HYPOTHESIS TESTING—WHAT COULD HAPPEN

Regarding making managerial decisions, the following table illustrates what could happen. Remember, you will never know for certain if H_0 is true or false because you are making an inference.

Your decision	H_0 is True	H_0 is False
Reject H_0	Type I Error	Correct Decision
Do Not Reject H_0	Correct Decision	Type II Error

In the above table, both Type I and Type II errors are just names given to the errors that can be made. They have no meaning in and of themselves and are only for communication convenience.

You choose the value of α for hypothesis testing: the smaller the value of α you choose for hypothesis testing, the smaller the chance of making a Type I error. At this point the only way to control for a Type II error is to take a larger sample!

> We know that a Type I error is rejecting a true null hypothesis H_0 and a Type II error is failure to reject a false H_0. What then is a Type III error? It is the probability of a decision-maker paying absolutely no attention whatsoever to Type I and Type II errors in hypothesis testing!!!—Dr. Gary Ramseyer

The Concept of POWER

The concept of statistical power is closely related to hypothesis testing. The POWER of any statistical procedure is defined as the probability that it will reject a false null hypothesis. Statistical power is inversely related to β, or the probability of making a Type II error. POWER $= 1 - \beta$

This definition applies to BOTH types of statistics. For a given sample size, a parametric statistics procedure is more powerful than an equivalent nonparametric procedure. However, a nonparametric statistics procedure can be as powerful as its equivalent parametric procedure by simply taking a larger sample. How much larger? Generally, a 10% increase in sample size will achieve equivalency. For example, in the first decision situation described above (the one about the two brands of tires), if a sample size of 100 achieved satisfactory results when using a t-test (a parametric procedure), the Mann-Whitney test (a nonparametric procedure) would require a sample size of 110 to be just as powerful.

Note that POWER has absolutely nothing to do with the data itself, or the scale upon which it was measured.

FREEBIE

All of the examples in this book are generated by the spreadsheet (computer program that is very easy to use) "OpenOffice," a Microsoft Excel look like, work like that has a good basic statistics capability built into it. Please do not "be put off" if you have never used a spreadsheet before. It is quite easy to use, and the examples in this book will have you moving around the spreadsheet in no time. See the Q & D Tutorial for instructions on how to download it at no cost to you, install it, and get started.

SUMMARY

The above introduction provides a justification for why you put forth the effort, time, and expense of studying the subject called "nonparametric statistics." The purpose of this Introduction is to get you to think about how nonparametric statistics can be used to support the decision-making process, and to start thinking statistically. It has introduced some terms and definitions that you will see (and use) again.

You have been introduced to:

- The generation of information to support decision making—Why you learn nonparametric statistics
- The definition of nonparametric statistics and some everyday uses of statistics
- Four facts about nonparametric statistics
- Decision-making situations and some examples
- Two assumptions that must be met in order for nonparametric statistics to work properly
- The concept of inference and that you never prove anything with nonparametric statistics

- The concept of hypothesis testing, how it is tied to inference, and how it supports decision making
- The concept of POWER of any statistical procedure.
- A free spreadsheet program OpenOffice that has some nonparametric statistics capability built-in and can easily perform nonparametric statistical procedures

4
Four Levels (Scales) of Data Measurement

It's All About Information

One accurate measurement is worth a thousand expert opinion.
Grace Hopper (1906–1992) Navy Rear Admiral, computer pioneer, coined the term "computer bug"

There are four (widely recognized) levels or scales of data measurement. Each level or scale is based upon the amount of information inherent in data measured on that level/scale.

- NOMINAL—the specification of data into categories and/or identification without any consideration of order or structure. Only differences are indicated.
  Consider race car numbers for the 2017 season. In NASCAR, the numbers 5, 24, 48, and 88 can all be categorized as coming from Hendrick Motorsports, and the numbers 11, 18, 19, and 20 can all be categorized as coming from Joe Gibbs Racing. The numbers also identify the drivers. But it would be a mistake to consider car 48 to be twice as good or fast as car 24, or that car 20 gets almost twice the attention as does car 11. Numerically, the implications are there, but the numbers only identify.

  Consider all NFL football teams and jersey numbers. Is a person with jersey number 70 seven times more valuable or important than a person with a jersey number of 10? NO! All the jersey numbers do is identify and categorize the players (lineman or back on offense).

  This is the lowest measurement scale with respect to information. An appropriate nonparametric statistics procedure is required if valid, useful decision support information is to be produced. The chart below provides some appropriate nonparametric statistics procedures.

W. Beatty, *Decision Support Using Nonparametric Statistics*, SpringerBriefs in Statistics,
https://doi.org/10.1007/978-3-319-68264-8_4

Binomial data can be considered in the nominal measurement category since there are only two classifications—success or failure, yes or no, 1 or 0. The two classifications only identify into which category the observation belongs.

- ORDINAL—this is simply a ranking. Differences and direction of differences are indicated. For example, consider the most well known ordinal scale—the 10 scale used by men to rank women's beauty (and women to rank men's handsomeness). A 1 ranking means extreme lack of beauty, while a 10 means extreme beauty. There can be agreement as to beauty—a higher raking means greater beauty. But there are two "problems" with this scale of measurement that preclude valid parametric analysis. First, there is absolutely no consistency, no objectivity of rank assignment. One man may rank a woman as an 8, while his friend ranks the same woman as a 6. Second, there is no measure of distance between points on the scale. A woman with a rank of 8 does not mean that she is twice as beautiful as a woman that receives a 4, only that she is more beautiful. Also, the distance between a rank of 8 and 4 may be vastly different between raters.
  Regarding information, data measured on the ordinal scale provides an evaluation (nominal) and preference (ordinal). Data measured on this measurement scale provides more information than data measured on the nominal scale.

  > ALL men think they rank as a 10 regardless of the number of teeth they have! ;)

  Data measured on the ordinal scale have more information than data measured on the nominal scale. Appropriate nonparametric statistics procedures are required if valid, useful decision support information is to be produced. Additionally, ANY procedure appropriate to analyze nominal data can be used to analyze ordinal data since all the information in ordinal data is present if it is analyzed as nominal data.

  And, regarding Grace Hopper, she would rank a 1 on beauty, but a 10 on desirability.

- INTERVAL—this scale has an **arbitrary** zero point, but (some say) it has equal distance between scale points, so the amount of difference (as well as its direction) can be determined. The most well known Interval scale—the Likert scale (invented by educator and psychologist Rensis Likert)—most commonly has five observation points: Strongly Agree, Agree, Undecided, Disagree, and Strongly Disagree. The numbers 1–5 are (usually arbitrarily) assigned to each scale point. But there is no zero value, so what does the value 0 mean? Even assigning the values 2, 1, 0, −1, −2 does not answer the question because 0 is arbitrarily assigned. Further, this scale assumes equal distance between scale points. Is the distance (in terms of feelings or opinion) between, say, 1 and 3 the same for every respondent? What about the distance between 1 and −1? For the above reasons, data measured on the Likert scale SHOULD be analyzed as ordinal data. It is mentioned here because some marketers maintain that data collected using the Likert scale has ratio (the next scale) scale characteristics.

The other most well known Interval scales are Celsius and Fahrenheit temperature scales. Both have consistent distances between scale points. For example, a certain specific amount of energy is required to change, say, a litre of water one point (or any number of points) on either scale. The scale points are consistent.

But, the zero scale point is defined as the temperature at which water freezes, or solidifies. That's fine, but why water? Why not oxygen or lead (or any other element)? For this reason data measured on the interval scale should be analyzed using nonparametric statistics procedures. However, because scale point distances are consistent, you can "get away with" considering data to have Ratio scale characteristics, and can be analyzed using parametric, or "regular," statistics procedures.

Consider also that 0 degrees C or F does not mean the absence of temperature, and therefore negative or minus temperature points are possible. Negative values are possible with data measured on the Interval scale, thus eliminating consistency, a major problem with using "regular" statistical procedures to analyze data measured on the Interval scale. For example, a person may have a negative attitude toward a product and assign a −2 on the Likert scale. Here the negative value can be interpreted. But how is a negative production rate for a plant interpreted? Be sure to be able to justify yourself if you use "regular" statistics analysis procedures to analyze data measured on the Interval scale.

What about the Kelvin temperature scale, absolute zero? Absolute zero occurs when all molecular motion ceases, an absolute, measurable point. So temperature measured on the Kelvin scale can be considered to be Ratio (the next scale) in nature.

- RATIO—this scale has an **absolute** zero point. Data measured on this scale is considered to have the most information. Zero means zero, or the complete absence of whatever you are measuring. For example, a production rate of 0 means that the factory is producing absolutely nothing, is idle. Negative or minus production rates (that make any sense) are not possible due to the absolute zero point. The name "ratio" also has meaning. For example, a person earning $10/h earns exactly half what a person earning $20/h, a 1 to 2 ratio. A factory that produces 30,000 computer chips per hour has a production rate three times that of a factory that produces 10,000 computer chips per hour, a 3 to 1 ratio. Further, doubling (or trebling, or...) the earnings (or production rate) does not alter the ratio. The same cannot be said for any other measurement scale. This is an important point, because the use of "regular" parametric analysis on ANY data not having the ratio property produces uninterpretable garbage.

  ANY analysis is appropriate for data measured on this scale. Even the nonparametric statistics analysis tools introduced in this book are appropriate.

Regarding information, as an example, examine the same data measured on the ordinal and ratio scales. Consider the hourly wage rates for two identified workers: one making $20/h and one making $40/h. Measured using the ordinal scale, you can tell only which one makes more per hour than the other: one piece of information. The same data measured on the ratio scale permits you to determine which worker makes more per hour and the ratio of how much more one worker makes than the other: an additional piece of information.

Remember, statistical analysis does not produce information, it only summarizes it. Analysis cannot create information, cannot produce accurate or useful decision support information if the data itself does not contain the information.

So what? What do these four scales have to do with statistical procedure selection? Data measured on the first two scales, nominal and ordinal, **MUST** be analyzed using appropriate nonparametric statistics procedures if useful decision support information is to be produced.

Technically, data measured on the Interval scale should be analyzed using nonparametric statistics procedures. But analysis using parametric, or "regular," statistics procedures are often used because the two scales (interval and ratio) are very similar with respect to information. Just be able to justify your analysis

Data measured on the ratio scale can be analyzed using ANY statistical procedure, but is most often analyzed using "regular" or parametric statistics procedures since they yield the most decision support information.

It is incumbent upon YOU, the decision maker, to determine which scale of measurement was used to measure your data. You should always use an analysis procedure appropriate for the data measurement level or scale. You can use nonparametric statistics procedures to analyze ratio data, but nonparametric analysis, for a given sample, are not as powerful, do not produce as much decision support information. Studies show that most nonparametric analyses require about 10% more data to be as powerful (produce the same decision support information) as analysis using parametric analysis procedures.

When designing your research and data collection process, data measurement scales should be a consideration when generating information in support of the decision you have to make. Businesses generate vast quantities of data that can be measured on the ratio scale, hence the first statistics courses focus upon "regular" parametric statistics and analysis.

But business managers (decision makers) are often confronted with data measured using nominal, ordinal, or interval scales. So, having nonparametric statistics procedures can come in handy, and improve decision making prowess.

Bottom Line: you cannot force information creation through analysis if the information is not there in the first place. Use nonparametric statistics procedures if you have data measured on nominal or ordinal (or interval) scales. If you are fortunate enough to have data measured on the ratio scale, you are free to choose any procedure you wish.

Figure 4.1 illustrates what analyses should be conducted based on measurement scales.

Type of Data and Applicable Statistical Tests		
Measurement Scale	Analysis	Typical Analysis Tools
Nominal	Nonparametric	Runs Test (chapter 7), Sign Test (chapter 7, 8), Chi-Square Contingency Table (chapter 8, 9), McNemar Test (chapter 8)
Ordinal	Nonparametric	Wilcoxon Signed Rank Test (chapter 8), Mann-Whitney U Test (chapter 8), Kruskall-Wallis Test (chapter 9), Spearman's Rank (chapter 10)
Interval	Usually Nonparametric, but can be Parametric (see caution in text)	Wilcoxon Signed Rank Test (chapter 8), Mann-Whitney U Test (chapter 8), Kruskall-Wallis Test (chapter 9), Spearman's Rank (chapter 10), t-test, ANOVA
Ratio	Any applicable nonparametric procedure, but usually Parametric procedures	t-test, ANOVA (note: t-test and ANOVA are "regular" (parametric) procedures that are NOT addressed in this book)

Figure 4.1: Four levels of measurement and appropriate analysis

5
Variables, Random Variables, and Variable Characteristics, Sampling

VARIABLES AND RANDOM VARIABLES

A **variable** is a name or symbol for the identification of any unique population characteristic that is subjected to statistical analysis. A **random variable** has its value determined by a random chance event, the conduct of a statistical experiment, such as sampling. A variable and random variable are not the same thing. A variable identifies, but a random variable assumes a value.

> All dogs are animals. All cats are animals. Therefore, all dogs are cats.—Keith Sullivan

A single study can have more than one variable, and, therefore, more than one random variable. A random variable can assume any value, but its values are usually defined by the analysis being conducted. For example, if you are analyzing heights of college students (the variable), you would expect the height random variable to assume values of between 48 and 84 in. You would not expect the height random variable to have a value of 24 or 120.

There are two kinds of variables and random variables: discrete and continuous.

> Statisticians do it continuously but discretely.—Chris Morton
>
> Two random variables were gossiping and thought they were discrete by whispering, but I heard their chatter continuously.—Dan Hayden

Discrete Variables are just what you would expect—discrete, or singular, measurements of outcomes. The value of the random variable produced

W. Beatty, *Decision Support Using Nonparametric Statistics*,
SpringerBriefs in Statistics,
https://doi.org/10.1007/978-3-319-68264-8_5

from a discrete variable must also be discrete. If a random variable can assume ANY value then it is said to be continuous. The value of the random variable produced from a continuous variable must also be continuous.

Here is an example for two dice being rolled. The integers 2 through 12 **MUST** appear, and it is not possible to get any other values, such as 5.5 or 13. In other words, the random variable value MUST be between 2 and 12 and the value MUST be an integer. The rolling of two dice yields a discrete random variable value.

The variable is called a **binomial variable** if its value can have only two values, typically defined as "success" and "failure." Note that any number of discrete variable values can occur so long as a value (or some combination of values) is considered a success. For example, the result of tossing a coin, with "heads" up defined as success, is a binomial random variable. Tossing a die, with getting any odd value between 1 and 6 defined as success, defines a binomial random variable. Tossing two dice, with getting an even value sum defined as success, defines a binomial random variable.

A numerical value assigned to a person's opinion is an example of a **Continuous Random Variable** because it can take on ANY value. For example, a person's opinion of, say, satisfaction with a product can be 50, or 55.5, or 60.54, or 70.547, or .... The opinion can be assigned any value. You are limited only by your imagination.

Note that a continuous random variable become discrete when its values are assigned to two (and only two) categories, a binomial random variable. For example, if you consider any positive opinion value as success, and any negative opinion value as failure, the variable and random variable become binomial.

Your definition of the variable and the values the random variable can assume determine whether it is continuous or discrete. For example, if you consider a person's opinion only in integer values between 1 and 10, then the variable "opinion" is discrete. A person cannot have more than one opinion, and that opinion must be expressed only with integer value. If opinion can be expressed in terms of decimal numbers, such as 7.4356..., the variable becomes continuous.

For BOTH kinds of variables, it is possible to calculate a test statistic (used for hypothesis testing) that can take on ANY value. For example, consider a situation where you solicit customers' opinions as integer values only. In almost all instances, the test statistic you calculate will have a continuous value. This fact, as it turns out, is just fine.

SAMPLING

One day some papers catch fire in a wastebasket in the Dean's office. Luckily, a physicist, a chemist, and a statistician happen to be nearby. Naturally, they rush in to help. The physicist whips out a slide rule and starts to calculate how much energy would have to be removed from the fire in order to stop the combustion. The chemist works on determining which reagent would have to be added to the fire to prevent oxidation. While they are doing this, the statistician is setting fires to all the other wastebaskets in the adjacent offices. "What are you doing?" the Dean demands. To which the statistician replies, "To solve a problem of this magnitude, you need a larger sample size."

Statistics Movies that Never Made it to the Big Screen: License To Sample

Sampling (drawing or taking a sample) is simply identifying and subsequently selecting for further analysis a "manageable" subset of the population of interest that you defined. How you define manageable will often determine the size of (number of members in) your sample. Manageable is usually defined in terms of time and/or money, but other factors, such as destructive analysis, can be considered. There is no minimum sample size for most statistical procedures. The "rule of thumb" is to take as large a sample as you can in order to reduce sampling error as much as possible.

The test statistic value is calculated from the data values of the observations drawn in the sample, one test statistic for each variable identified.

There is also the concept of **random sampling**. Random simply means that EVERY individual member of the population you defined has an equal chance of being included in your sample. Consider, for example, asking people's opinions. Let's assume that the store where you are asking about opinions concerning service is open 24 h per day. Sampling only from 8 AM until 4 PM would not yield a random sample since all people in other hours cannot be included in the sample. Sampling **bias** could let other factors slip by, such as a night manager being surly. Considering ALL opinions, regardless of time, can be considered a random sample.

It is **YOUR** responsibility to see that samples are taken randomly. There is no statistical test for this. Selective or non-random sampling virtually guarantees the statistical analysis will produce corrupted results. There is a saying with computers that is applicable here: GIGO, or Garbage In,

Garbage Out. If your sample is not taken randomly, it will not reflect reality, is garbage, and its analysis will not yield reliable results.

The concept of **independence** when sampling is also an important factor. Independence means that each sample member is not affected by any other sample member (with regard to inclusion in the sample, or its value). For example, you would not want to ask opinions from two (or more) people from the same family since opinions are often shared (not independent) among family members, and asking opinions of two (or more) family members may violate the independence principle since opinions are often influenced by other family members.

It is **YOUR** responsibility to see that samples are taken independently. There is no statistical test for this.

SAMPLING ERROR

Sampling error (often called estimation error) is the amount of inaccuracy in estimating a population parameter that is caused by using only a part of the population, the sample, rather than the entire population. When you take a sample from a population in order to calculate a test statistic, the test statistic value will never be the same as the value of the population parameter it estimates. There is no way to avoid this situation because a sample is never an exact duplication of the population from which it is taken. The difference between the population parameter and its estimate, the test statistic, is called the "sampling error." The sample is never an exact representation of the entire population, so the value of the test statistic calculated from the sample will never be exactly the same as the unknown population parameter. Sampling error cannot be eliminated. But it can be reduced by two methods.

First, the size of the sampling error can be reduced by taking random samples because random sampling eliminates the introduction of bias. And, identifying independent sample members also helps reduce bias.

Second, the larger the sample, the greater the chance the sample accurately reflects and is representative of the total population. The downside is that the larger the sample, the more expensive and/or time consuming it is to take. You need to find a balance between sample size and time/cost. Often the size of the sample is determined by how much time and money you have.

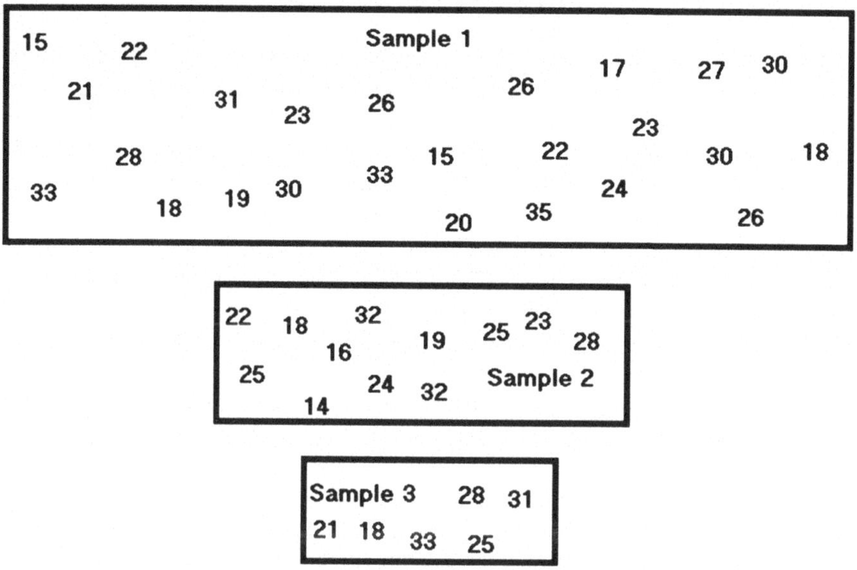

Figure 5.1: Sampling error

In Fig. 5.1, assume (for illustrative purposes) that these rectangles represent samples taken from some population. Let's say that the population is ALL the customer satisfaction opinion scores, and you want to estimate the median customer satisfaction score. (Higher scores mean greater satisfaction)

Rectangle **1** represents the first sample taken (from the entire population), and it has a sample size of 25, since there are 25 satisfaction scores in it. It is the most expensive and time consuming to conduct.
Rectangle **2** represents the second sample taken (from the same population), and it has a sample size of 12.
Rectangle **3** represents the third sample taken (again, from the same population), is the smallest, with a sample size of only 6. It is the cheapest and easiest to conduct.

Sample 1 (the largest) has the greatest chance of "abnormal" scores, such as having a very disgruntled customer, being offset by more typical satisfaction scores. Sample 3 (the smallest) has a much less chance of abnormalities being offset by typical satisfaction scores. As you can see, the larger the sample, the greater the chance that the sample is representative of the entire population and the better the resulting decision will be.

Usually only one sample is taken and the decision made is based on it, but for explanation purposes three samples were considered here.

MARGIN OF ERROR

The margin of error is a value (usually expressed as a ± [plus/minus] percentage) expressing the amount of sampling error that CAN occur from the results of sample analysis. A margin of error occurs when a population is incompletely sampled, so the margin of error is mostly concerned with **HOW** a sample is conducted. Consider this example. A polling company says its sampling method yields a ±2% margin of error. Its sample analysis says that 60% of city residents favor an issue. But the actual percentage could be as low as 58% (−2%) or as high as 62% (+2%). Then the polling company can express (with some degree of certainty, usually 95%) what the true population of all city residents thinks about the issue, that it is within the ± range.

Another polling company says its sampling method yields the same ±2% margin of error. Its sample says that 58% of city residents favor an issue.

So which one is correct? The answer is "Yes," that they both are. Such is the nature of variability, sampling, and sampling error. The margin of error decreases as the sample size increases. The only way to have the margin of error be zero, to not commit a sampling error is to have your sample size be ultimately large, to take a census.

A new government 10 year survey that cost in excess of $3,000,000,000, revealed that 3/4 of the people in America make up 75% of the population

SUMMARY

You learned that there are only two kinds of variables—continuous and discrete.

You have been introduced to:

- Variables and Random Variables
- Sampling
- The concept of Sampling Error and how to reduce it
- The concept of Margin of Error

6

Probability and the Binomial Probability Distribution

All probabilities are 50% ... either something happens, or it doesn't!

"Do you realize there is some probability that before you complete this sentence, you will be hoofed insensible by a wayward, miniature Mediterranean ass?" Stephen D. Unwin, *The Probability of God: a Simple Calculation that Proves the Ultimate Truth*
In other words, "Some probability" does not equate to high probability.

Statistics Movies that Never Made it to the Big Screen: The Wizard of Odds

Understanding How Probability (Chance) Works

Before going into ANY type of Statistics, it is imperative that you know something about probability, since much of Statistics is based on probabilities. Fortunately, probability is quite simple and easy to understand. And no, all probabilities are NOT 50%.

Did you hear the one about the statistician? Probably...

DEFINITIONS

Before we begin, here are some definitions with which you must be familiar. Why have these definitions? There are two reasons:
1. They aid communication. Remember, all this "statistics talk" is just a communication aid so that basic statistical concepts don't have to be re-explained.

W. Beatty, *Decision Support Using Nonparametric Statistics*, SpringerBriefs in Statistics,
https://doi.org/10.1007/978-3-319-68264-8_6

2. Absolute precision of communication is mandatory. It is necessary that you "say what you mean and mean what you say." To do otherwise will invalidate the analysis results due to miscommunication.

So, with that said, let's proceed with definitions:

- An **Experiment** is the description of a situation that leads to a result in which probability (or chance) is involved. For example, an experiment is conducted each time an opinion is ascertained or the tosing of a pair of dice. A single iteration of an experiment is called a **trial**.
- An **Event** is generated each time an experiment is conducted. For example, each asking of an opinion or each tossing of the dice constitutes an event.
- The **Event Space** is the enumeration of all possible outcomes of an event. For example, if a five point Likert Scale is used [extreme like, like, neutral, dislike, extreme dislike], the event space is 5, 4, 3, 2, or 1. No other outcomes are possible. The event space is 1 to 5. With the dice, the event space is an integer between 2 and 12.
- An **Outcome** is the measurable result of an event. For example, when an opinion is ascertained, the outcome, or result, or value, will be one of the five values on the Likert Scale. For the dice, the sum of the dots on the "up" sides is the outcome.
- A **Probability** is a measure of the likelihood that something specific (outcome) will occur—the next time the experiment occurs. For example, when an opinion is ascertained, the probability of the opinion being a specific outcome (for example, "like") is 1 in 5, or 20%, or 1 possible opinion result divided by 5—all possible opinion results. For the dice, there are 36 possible outcomes. So the probability of a specific outcome is simply the number of ways the outcome can occur divided by 36. There are six unique ways toss a 7, so the probability of tossing a 7 on the next toss is 6/36, or 0.16666..., or 16.6666%.

The notation of a specific outcome is: **P(outcome)**. Often the specific outcome is represented as A, so you often see the notation P(A).

An experiment, an event, and an outcome are NOT the same things. An event is the occurrence of an experiment. The outcome is the measurable result of the conduct of the experiment. An experiment is the description of a situation.

As you can see, these definitions are all interconnected, so you need to understand all five.

The "naming" of the outcome is what is known as a **variable**. The "value" of the outcome, what the trial generates, is what is known as a **random variable**.

USING THE DEFINITIONS

An event occurs each time an experiment (trial) is conducted. An outcome is generated each time an event occurs. The event space is a list of all possible outcomes of an event. A probability can be calculated for each outcome that results from a trial.

Consider this: an "asking opinions" example. The "asking opinions" situation defines an experiment.

- The action of the asking for an opinion is an event.
- The event space is the actual opinion that can be expressed (for simplicity, opinions can be only "favorable" or "unfavorable").
- The outcome is the actual opinion expressed.
- A probability of each specific outcome can be assigned when the next opinion is ascertained.

The variable is the "opinion expressed", while the random variable is the actual opinion (favorable or unfavorable). If "1" means favorable, and "0" means unfavorable, then the random variable (value 0 or 1) can be considered binomial.

SIMPLE EXAMPLE

Here is a simple example that illustrates all five definitions and introduces probability. Consider a coin toss, a binomial experiment. The coin is "fair," meaning that the probability is unchanged on each toss, it has no unusual weighting, and it has a head side and a tail side. You can ignore the possibility that the coin could land on its edge because the probability of that outcome is astronomically low. When the coin lands, one of the two sides will be visible, or up. Since there are only two sides to the coin, one of the two sides must be up. Therefore each side has a 50% probability (one divided by two) of being up. One of the two sides must be up and since each of the two sides has equal chance of being up, each side has a one in two chance of being up on the next toss. One divided by two yields the probability of each side being up, in his case 0.5, or 50%.

Each separate coin toss (trial) is an event with an observable outcome (one or the other side can be identified). Any time there are only two possible outcomes (as with the coin toss), the experiment is considered to be binomial.

> Try this simple test: flip a coin, over and over again, calling out "Heads" or "Tails" after each flip. Half the time people will ask you to please stop.—Jack Handey

CALCULATING PROBABILITY

You can, by using probability, determine in advance the likelihood of outcomes that can occur. You cannot determine precisely the next outcome, but you can determine the likelihood or probability of a specific outcome. You can determine what should happen if a "large" number of experiments or trials are conducted. The question then becomes: "What does 'large' mean?" More on that later.

The probability of each side being up can be calculated as:

the number of outcomes that can occur each time the experiment is conducted

divided by

the total number of possible outcomes that can possibly occur each time the event occurs (all possible outcomes enumerated by the event space)

For the coin toss, each conduct of the experiment yields a single outcome—one side of the coin must be up, only one outcome is possible. Since there are only two sides of the coin, the total enumeration of all possible outcomes is two. Each side has a 50% probability (chance) of being up since one divided by two yields 0.5, or 50%, or a one in two chance of being up. If "heads" is represented by A, then P(A) = 0.5.

It's that simple. ALL probability works the same way.

It is imperative that you define everything precisely. Calculating the probability for each possible outcome in the event space, then placing them in tabular form, is what is known as constructing a **probability distribution**.

ANOTHER EXAMPLE

Consider the toss of a single die. Each time the die is tossed is the conduct of an experiment. An event occurs and only one outcome is possible. The outcome of the event is defined as the number of dots on the face of the die that is up. Since there are six sides on a single die, with each side having a unique number of dots, from one to six dots must be on the side that is up. The event space is all the possible number of dots (outcome) that can be showing on the side that is up, or from one to six dots. Therefore, the event space is 1, 2, 3, 4, 5, and 6. The probability that any side will be up on the next toss is simply one divided by six, or 16.667%. Notation: If A is defined as "a particular side will be up," then P(A) = 0.166666667, or 16.667%. If A is defined as "any side will be up," then P(A) = 100%.

What is the probability of getting a 7 on the next toss? Since the event space for a single die has been defined as from 1 to 6, P(7) = 0. There are two important statistics truisms represented here:

- The probability of obtaining ANY result not in the event space is always zero.
- The SUM of all the probabilities for each possible outcome in the event space must be 100% (or very near 100%, a difference due only to rounding).

Let's examine another example. Assume that you are ascertaining service level opinions of entering first time customers at a local WalMart. Assume that the only opinion responses you allow are: Good, Neutral, and Not Good. If all three choices are equally, represented, what is the probability that the next entering customer will express "Good" as his/her opinion? "Good" is a possible response, and there are three possible responses, so P(Good) = 0.3333.... What is the probability that "Excellent" will be expressed? Since "Excellent" is not an opinion choice, so P(Excellent) = 0.

The manager of the WalMart has worked diligently on service, so his opinion is that the three choices have a 75%, 20%, and 5% chance of occurrence. If he is correct, the probability of the next customer responding "good" is 0.75, or 75%. But we get ahead of ourselves here. Patience!

The service level opinions will (hopefully) be modified by the actual service levels received, and opinion probabilities are changed. So, the probability of opinion responses from exiting customers is not the same thing as for first time entering customers. Customers' opinions may be changed while in the WalMart. Ascertaining opinions of entering customers is independent with respect to response, but ascertaining opinions of exiting customers is not independent since customer opinions can be affected by service. The analysis of "change in attitude" (entering versus exiting) is addressed in Chap. 8.

BOTH independence and the random property must be present if valid, useful information is to be generated.

MUTUALLY EXCLUSIVE, COLLECTIVELY EXHAUSTIVE, INDEPENDENT—three very important statistics concepts

In both examples above (as with ALL probability), the events are mutually exclusive. Each event and its resulting outcome is unique. It is not possible for more than one outcome to occur each time the experiment is conducted. "Unique" in this case means singular. The event space is collectively exhaustive. Every possible outcome is enumerated, is considered. Each time the experiment is conducted, the result is independent of, or not affected by, the conducting of previous experiments.

There are no tests for these three concepts, so it is YOUR responsibility to see that they are met.

ANOTHER EXAMPLE

With a single die, what is the probability of tossing an even number on the next toss? The event space is [the tossing of an even number] or [tossing an odd number]. The outcome of the event is 2, 4, or 6 (even numbers) or 1, 3, or 5 (odd numbers). The event space is either even or odd, one of two results. The probability of tossing an even number is: 1 divided by 2, or 0.5, or 50%. If A is defined as "tossing an even number," then P(A) = 0.5.

YET ANOTHER EXAMPLE

Consider two dice being tossed. The result of the event is defined as the sum of the dots showing on the sides that are up—for that toss. The event space is 1 and 1, (snake eyes), 1 and 2, 1 and 3 ... 4 and 6, 5 and 6, or 6 and 6 (boxcar). One of these outcomes must occur each time the two dice are tossed. The event space has a total of 36 unique possibilities.

There is only one way to get snake eyes (1 and 1 for a sum of 2), so on the next toss the probability of getting snake eyes is 1 in 36, or 2.7777777...%. Calculating the probability of a boxcar is just like calculating the probability of snake eyes.

Consider the probability of getting a six on the next toss. There are five different ways to get a six—1 and 5, 2 and 4, 3 and 3, 4 and 2, and 5 and 1. Remember, there are two dice, so order enters into play here. Think of the two dice as being different colors: red for the first die, and green for the second die. You must look at the outcome of the first die, the red one, then the second die, the green one, and then sum the two outcomes from the two dice into a single outcome. So the probability of tossing a six is 5/36, or 13.8888888...%. There are five ways to roll a six: 1 (red) and 5 (green), 2 (red) and 4 (green), 3 (red) and 3 (green), 4 (red) and 2 (green), 5 (red) and 1 (green), and there are thirty six possible outcomes.

Now consider the probability of getting a seven on the next toss. There are six different ways to get a seven—1 and 6, 2 and 5, 3 and 4, 4 and 3, 5 and 2, or 6 and 1 (remember, there are two dice). Since there are six ways to get a seven, the probability of tossing a seven is 6/36, or 16.6666666...%.

Obviously, the concept of "**on the next toss**" is important. Here we are calculating what may happen only for the next dice toss. Calculating a probability of a specific outcome for the next event (something that is impossible to do) is quite different from calculating a probability distribution—what will probably happen if a "large" number of tosses occur. (Remember, in "craps" you are always wagering on the next toss)

YET ANOTHER EXAMPLE

Consider again two dice, but change your perspective this time. Two dice are tossed, and the result of the event is defined as the sum of the dots showing on the sides that are up without regard to which die was first or second: this time you are interested only in the sum, not in dice order. Even though there are now eleven possible outcomes: 2, 3, 4, 5, 6, 7, 8, 9, 10, 11, and 12, there are still thirty-six ways to achieve these outcomes. So this example is exactly like the one above. Whether you consider dice order, and then sum them, or consider only the dice sum, the outcome probability is still the same.

YET ANOTHER EXAMPLE
Sampling With and Without Replacement

Consider now that you are blindly drawing black and white marbles from a covered jar (sampling). Let us say that there are 12 black marbles in the jar and 8 white marbles in the jar. Each time you conduct the experiment, you draw only one marble from the jar.

The first time you conduct the experiment, what is the probability (chance) that you will draw a (sample of a) black marble? Since 12 of the 20 marbles are black, there is a 12 in 20 chance (60%) of drawing a black marble.

Now you conduct the experiment again. There are two different ways to continue conducting the experiment.

You can replace the marble you just drew out, therefore restoring the count to 12 black marbles and 8 white marbles, and restoring the probabilities to exactly what they were the first time you conducted the experiment. By returning the marble drawn and returning probabilities to what they were, you are **sampling with replacement**.

If, however, you do not replace the marble drawn in the first experiment, the probability of drawing a black marble this time is affected by the result of the first time you conducted the experiment. In this case, the two experiments—the one you just conducted and the one you are about to conduct—are not independent. If, when the first experiment was conducted, you drew a black marble, there are, before the second experiment is conducted, 11 black marbles and 8 white marbles from which to draw (sample). Now the probability of drawing a black marble this time has changed from the first time the experiment was conducted. This situation is called **sampling without replacement**.

The size of your population should be a consideration when deciding upon sampling with or without replacement. When sampling from a small population, sampling without replacement can affect probability. If, however, you are sampling from a large population, statistics assumes that sampling does not change probabilities enough to be of consequence.

What defines "small" and "large?" The answer is "YES." If the change of probability can affect the decision, then it is better to sample with replacement and consider the population to be small. If, however, sampling causes the probability to changes very little (has no effect on the decision) then the population can be considered to be large and warrant sampling without replacement. Say, for example, you have 20,000 marbles, 12,000 black and 8000 white. On your first draw, the probability of drawing a black marble is exactly the same as with 20 marbles example above. Now let's say that you do not return the marble drawn. Does that change the probability of drawing a black marble next time? Technically, yes. But the probability change of drawing a black marble is so small that it can be considered negligible and can be ignored.

How about with 10,000 marbles? How about with 5000 marbles? How about with 1000 marbles? How about with 500 marbles? How about with 100 marbles? Hopefully, you can begin to see that "small" and "large" have no specific definitions.

So, ultimately, there is no precise definition of "large." When sampling, "large" means "unaffected probability." When calculating probability (as with dice tosses), "large" means "enough trials so that what should happen, does."

Why is all of this important? What does drawing marbles or tossing dice have to do with making business decisions? Think about this scenario: replace marbles (or dice) with people and their opinions. The size of the population you study, how you calculate probabilities, and how you sample from it can definitely have an effect on your business decision.

SOME "COMMON" PROBABILITIES

We are all familiar with these (and other) "probabilities."

- Winning Number at Bingo: Needing one more number to win at Bingo is called a "wait." For example, let's say that you have covered B-3, I-16, N-40, and G-59. All you need to win is O-73. You are waiting for O-73 to be called so you can win. The probability that O-73 will be called next can be precisely calculated. But the problem occurs when several other people in the same Bingo game also have a "wait" different from yours. So, the probability that you will win must also consider the other "waits" as well. The probability that you will win can still be exactly calculated, but this environment takes you into an entirely, much more advanced type of statistics.

- Batting Averages: There are just too many factors involved to turn batting into a precise science. A higher batting average may be indicative of one batter's ability vis-à-vis another batter, but a higher batting average cannot offer a precise prediction of what will occur the NEXT time a batter appears. But ... Probability gives a baseball team manager some degree of confidence in his decision making. If Dan Uggla, a right-handed batter, currently has a rather low overall batting average, but 7 for 10 lifetime against Stephen Strassberg, the current pitcher who is right-handed, the manager isn't going to pinch hit for Uggla with a left handed batter because a 0.700 (7 for 10) batting average is very high, so there is a high probability (though no guarantee) that Uggla will get a hit against Strassberg. However, many years of baseball records show that right-handed batters don't do as well against right-handed pitchers (and vice versa). Probabilities of getting a hit play a key role assisting the decision-making process.

- Winning the Lottery: What is the probability of forecasting the winning lottery number? There are some facts about ALL lotteries that you must remember. First, ALL lotteries keep some portion of sales proceeds (usually half). Second, the probability of winning is directly proportional to the number of ticket holders. More lottery players means a larger money payoff, but it also means that your probability of winning is diminished. Third, there is absolutely no way that statistics

can **precisely** predict what will happen next, what the precise winning lottery number will be. Statistics can help you with general future probabilities. The problem here, of course, is that lotteries don't pay based on general results.
Besides, don't you think that if I could, I would have already won the lottery and already have sold the secret to other lottery players?

THE BINOMIAL PROBABILITY DISTRIBUTION
A DISCRETE VARIABLE PROBABILITY DISTRIBUTION

The most common discrete variable probability distribution is the "Binomial" probability distribution. So, this book examines the Binomial probability distribution and uses it.

Because there are only two possible outcomes (0/1, success/failure, H/T, M/F, Like/Don't Like, etc.), Binomial data is considered to be nominal in nature.

Consider the tossing of two coins. The event space is: HH, HT, TH, or TT. The probability distribution (Fig. 6.1) is the expression of the event space in terms of what can happen. "But wait!" you say. There are four possible outcomes here—definitely not Binomial. Patience, patience!

In this example, the probability of getting two heads (or two tails) is 1/4. The probability of getting a head AND tail is 2/4, since there are two ways to get a head and tail. So, the probability distribution when tossing two coins is:

0 Heads (T,T)	1/4 or 25%
1 Head (T,H or H,T - two ways)	2/4 or 50%
2 Heads (H,H)	1/4 or 25%
Probability Sum	100%

Note that the probability distribution for "Tails" is the same as for "Heads."

0 Tails (H,H)	1/4 or 25%
1 Tail (T,H or H,T)	2/4 or 50%
2 Tails (T,T)	1/4 or 25%
Probability Sum	100%

Figure 6.1: Two coin probability distribution

These tables, which are probability distributions, expresses what will probably happen as a result of a "large" number of coin tosses, or events, occur. It cannot and does not express the probability of the outcome of the next coin toss. What does "large" mean? The answer is "YES"!

Note that this probability distribution applies ONLY to the probability of getting "Heads" and "Tails" for the tossing of two coins. This means that every event space has its own probability distribution.

Note also that the sum of all probabilities must be (very, very near, off only because of rounding) 100%.

Here's where the "Binomial" part comes in. If "success" is defined as getting two heads (OR two tails, OR no heads, OR no tails), then P(success) = 25% on the next flip of two coins. If "success" is getting one head (or tail), then P(success) is 50%. The definition of success defines its probability. Notice also that "**OR**" means that only one definition of success is possible for each trial.

EXAMPLES

OpenOffice has a binomial probability distribution function: BINOMDIST It has four arguments separated by semi-colons (;). For example, if a coin with (P(Success)=0.5) is tossed 25 times (trials—second argument), what is the probability that exactly (non-cumulative) 10 (first argument) tosses will be "heads"—successful? The P(success) for each toss is 50%, or 0.5 (the third function argument). The fourth function argument must be 0 in order for single probabilities to be reported, a property called non-cumulative. The BINOMDIST function is illustrated in Fig. 6.2. So, the probability of getting exactly 10 "heads" (the first function argument—see the "Input" line) in 25 coin tosses (the second function argument) is 0.0974166..., or 9.74166...%. Similarly, theprobability of getting exactly 5 heads (successes) in 25 tosses is 0.158...%. And the probability of getting exactly 20 heads in 25 tosses is 0.158...%

Now, say the probability for the coin to come up "heads" is 80% (P(Success)=0.8, the third argument) Again, calculate probabilities for 5, 10, and 20 successes in 25 tosses (trials). The results are illustrated in Fig. 6.3. The probability of getting 5 or 10 heads is very, very low, while the probability of getting 20 heads is 19.602...%. By the way, cell C7 contains 5, 10, or 25, the number of successes for which a probability is desired. (Note that the desired number of successes is in Cell C7)

Figure 6.4 illustrates the probabilities resulting from tossing two dice—a probability distribution.

Consider the probability of tossing a nine if two dice are tossed 100 times. P(Success) is 11.1% (from Fig. 6.4) each time the dice are tossed. So, what

For the experiment of tossing a coin (that has a probability of being "heads" 50% (0.5) of the time) 25 times, what is the probability of getting heads a specified number (no. successes) of times?
The function BINOMDIST answers the question.

No. Successes	Probability	
10	0.097416639	Prob of 10 heads in 25 flips (=BINOMDIST(10;25;0.5;0)
5	0.001583397	Prob of 5 heads in 25 flips (=BINOMDIST(5;25;0.5;0)
20	0.001583397	Prob of 20 heads in 25 flips (=BINOMDIST(20;25;0.5;0)

Figure 6.2: OpenOffice BINOMDIST function

	A	B	C	D	E	F	G	H
17								
18				This time there's an 80% (0.8) chance of getting "heads."				
19				Note also that the no. successes is in cell addresses.				
20	No. Successes		Probability					
21	10		1.15009E-05	Prob of 10 heads in 25 flips (=BINOMDIST(A21;25;0.8;0)				
22								
23	5		1.82553E-10	Prob of 5 heads in 25 flips (=BINOMDIST(A23;25;0.8;0)				
24								
25	20		0.196015103	Prob of 20 heads in 25 flips (=BINOMDIST(A25;25;0.8;0)				
26								

Figure 6.3: OpenOffice BINOMDIST function

Dice Sum	Probability
2	2.8%
3	5.5%
4	8.3%
5	11.1%
6	13.9%
7	16.7%
8	13.9%
9	11.1%
10	8.3%
11	5.5%
12	2.8%
Probability Sum	(very near) 100%

Figure 6.4: Probability distribution for two dice

is the probability of getting exactly 10 nines? The probability is illustrated in Fig. 6.5.

Note the argument values. Cell C3 has a 10 (or a 20) in it—the definition of the number of successes. There are 100 trials. The probability of success is 0.111 (the probability of tossing a 9 [from the probability distribution in Fig. 6.4]). The final argument value is 0. This is also the probability for tossing fives. You should be able to see why.

	A	B	C	D	E	F	G	H
28								
29				This time we're rolling two dice. The probability distribution is illustrated above. As can be seen, there is an 11.1% probability of rolling a "9."				
30								
31								
32	No. Successes		Probability					
33	10		0.12379215	Prob of 10 "9"s in 25 rolls (=BINOMDIST(A33;25;0.111;0)				
34								
35	20		0.003529814	Prob of 20 "9"s in 25 rolls (=BINOMDIST(A35;25;0.111;0)				
36								

Figure 6.5: BINOMDIST function in use

So, the probability of getting exactly 10 nines in 100 tosses is 12.379...% (note that cell A33 has a 10 in it). And, the probability of getting exactly 20 nines is 0.00352...., or 0.352%. (P(9)=0.111, the third argument)

SUMMARY

In each of the examples presented, you learned about probability and how to calculate it, along with events, event spaces, outcomes, and experiments (trials).

You have been introduced to:

- The concept of an experiment and some definitions that you must know
- How to calculate probabilities
- The concepts of Mutually Exclusive, Collectively Exhaustive, and Independence
- The concepts of sampling with and without replace
- The concept of the Binomial Probability Distribution and how to calculate, using OpenOffice, a specific binomial probability

---

EXERCISES

1. In a bag are four green marbles, two red marbles, and six blue marbles. After each draw, all marbles are replaced in the bag. What is the probability of:
   (a) drawing out a green marble? 4/12
   (b) drawing out, in a single draw, both red marbles? 2/12
   (c) drawing out a blue marble? 6/12
   (d) not drawing out a blue marble? 6/12
   (e) drawing out a black marble? 0/12
2. Three of five city council members are to be selected to serve on a special planning committee. Mr. Grey is one of the five council members. What is the probability that Grey will be on the committee? 3/5
3. One card is chosen at random from a deck of 52 playing cards. What is the probability that the one card chosen will be:
   (a) an ace or a deuce? 8/52
   (b) a club? 13/52 or 1/4
4. Use the OpenOffice BINOMDIST function to find:
   (a) the probability of getting exactly 50 heads in 100 coin tosses, assuming heads has a 50% probability of being up.
   (b) the probability of getting exactly 50 heads in 100 coin tosses, assuming heads has a 70% probability of being up.
   (c) the probability of getting exactly 50 sevens in 100 rolls of two dice.
   (d) the probability of getting exactly 50 sevens in 100 rolls of a single die.
   (e) A fair coin (50% probability of either heads or tails being up) is tossed three times. What is the probability of obtaining one tail and two heads?

7

One Population Hypothesis Testing

The only relevant test of the validity of a hypothesis is comparison of its predictions with experience.
Milton Friedman (1912–2006) Economist and Statistician

I don't know why people are so negative about statistics and statisticians. I'm only a first-year student, and statistics has already taught me everything I need to know about life - always Proceed with Caution and Reject H_0!—Priscilla Mok, statistics student

NOMINAL DATA

THE RUNS TEST

The Runs Test addresses only nominal data, only two data categories, and only whether or not the sample observed was drawn from a population that is generated randomly.

An examination of runs can assist in deciding upon the randomness or nonrandomness of what is actually observed. Runs can be used to formulate tests of hypotheses regarding the randomness of the population from which the sample was drawn.

H_0: the number of runs observed is due to chance, that nothing is influencing the generation of runs,
H_0: number of runs = what is expected from a chance or random process
H_1: the number of runs observed is due to some factor other than chance, that something is influencing the generation of runs,
H_1: number of runs ≠ what is expected from a chance or random process

W. Beatty, *Decision Support Using Nonparametric Statistics*,
SpringerBriefs in Statistics,
https://doi.org/10.1007/978-3-319-68264-8_7

A run is defined as a simply a series (1 or more) of two differing values. For example, consider tossing a coin 26 times (the sample size: n=26) with this result: ("H" = heads, "T" = tails)

H H H T T T H H T T H H H T T T H H T T H H T T T T

In the example above, there are 10 runs (each run is underlined), which is the observed number of runs. Note that a run can be of any length: from 1 to ∞.

The "H" is arbitrarily changed to a 1 and a "T" is changed to a 0. Then,
the number of "1" ("H") observed (n1), which in this example = 12
the number of "0" ("T") observed (n2), which in this example = 14

So, we need to calculate:
the expected number of runs, often called R_{bar},
and the variance of the expected number of runs, often called $S^2{}_{Rbar}$

That is all that is needed to be calculated.
Using the formulae presented below, the expected number of runs (R_{bar}), and the variance of the expected number of runs ($S^2{}_{Rbar}$) can be calculated.

Expected number of runs = R_{bar} = (2* n1* n2)/((n1+ n2)+1)
where: n1 is the number of "1"s ("H") observed, and n2 is the number of "0"s ("T") observed.

The variance of the Expected number of runs $S^2{}_{Rbar}$ = ((2*n1*n2)*(2*n1*n2−n1−n2))/((n1* n2)^2)+(n1+n2−1))
The standard deviation (needed to calculate Z_{calc}) is simply the square root of the variance.

The above formulae appears formidable, so thank goodness for the "COPY" command, and that the RUNS spreadsheets can be downloaded. In other words, the hard work has already been done for you!

So, the test statistic, conveniently named Z_{calc}, is [the absolute value of]: (observed number of runs − R_{bar}) / $\sqrt{variance}$, [where "variance" = σ^2].

Also, Z_{calc} = [(observed number of runs − R_{bar})/σ] Note: this is the way it is calculated in the spreadsheets.

The test statistic is approximately normally distributed, so you can make good decisions by using the OpenOffice Normal Distribution function NORMINV.

WARNING: in order to use the OpenOffice NORMINV function properly (two-tailed test), the value of α (typically 0.05) MUST be halved, then subtracted from 1. That value (typically 0.975) is entered as the first argument into the NORMINV function ($1 - \alpha/2$;). The second and third arguments are ALWAYS 0 and 1 (the mean and standard deviation of the normal probability distribution are 0 and 1).

You can, alternately, use the NORMSINV function, where the second and third arguments (0 and 1) are assumed. All else is the same. Both functions return the same value.

Here is how you find the appropriate value by using the NORMINV function to get the "critical" Z value.

Typically, $\alpha = 0.05$ (you want to be [at least] 95% confident that you are correct regarding your decision about H_0), so first divide α by 2, yielding 0.025. Subtract that from 1, yielding 0.9750, is the first argument value entered in the NORMINV function (the second and third arguments are ALWAYS 0 and 1). The value returned is 1.95996... (or 1.96). This is your "critical" Z value.

Use this procedure to find the "critical" Z value when conducting any test (in this book) when using the NORMINV function.

The reason you must use this procedure is because the NORMINV function is, by design, a one-tailed function not suitable for direct use (because the OpenOffice programmers said so!)

Now calculate the test statistic (Z_{calc}) using the formula above: this is how many standard deviations what was actually observed (in your sample) is from the expected (hypothesized) number of runs (R_{Bar}). This quantity is called your calculated Z score or value. Why Z? Why not? It has to be called something. Z_{calc}, your calculated Z score, is actually the number of standard deviations what you observed differs from what you expected (hypothesized).

Now compare your calculated Z score with the Z returned by NORMINV (often called the critical Z, or Z_{crit}—note that "critical" is just a communication convenience). IF your calculated Z score is greater than the "critical" Z, REJECT H_0 and make your managerial decision based upon some other criterion (usually what you observed since it is available and you have nothing else upon which to go).

IF your calculated Z score is NOT greater than the "critical" Z, fail to reject H_0 (which is NOT the same as accepting H_0) and make your managerial decision as if H_0 is correct. (Note that H_0 was not accepted, or proven correct, or proven to be true, just not rejected.)

NOTE: if H_0 is correct (something you will never know for certain), then Z_{calc} will be "close" to zero. So, comparing Z_{calc} to Z_{crit} is simply the "statistics" way of comparing what you observed from the sample to what you hypothesized.

THAT'S ALL THERE IS TO IT!

In the coin toss example above, there are 10 runs (observed), the expected number of runs (R_{bar}) is 13.923, the standard deviation (square root of variance) of the expected number of runs ($\sqrt{S^2}_{Rbar}$) is 2.482, so the Z_{calc} score is -1.580
or $Z_{calc} = (10 - 13.923/2.482)$

The minus sign ($-$) indicates only that the number of runs observed is less than the number expected. So compare the absolute value of your calculated Z (1.580) score against your "critical" Z (1.95996). In this example, 1.580 is not greater than 1.95996 (the "critical" Z value returned by NORMINV), so DO NOT reject H_0. Your managerial decision can now be based what you observed as having been produced randomly since you could not, based on sample evidence, reject the idea of random coin flip procedure.

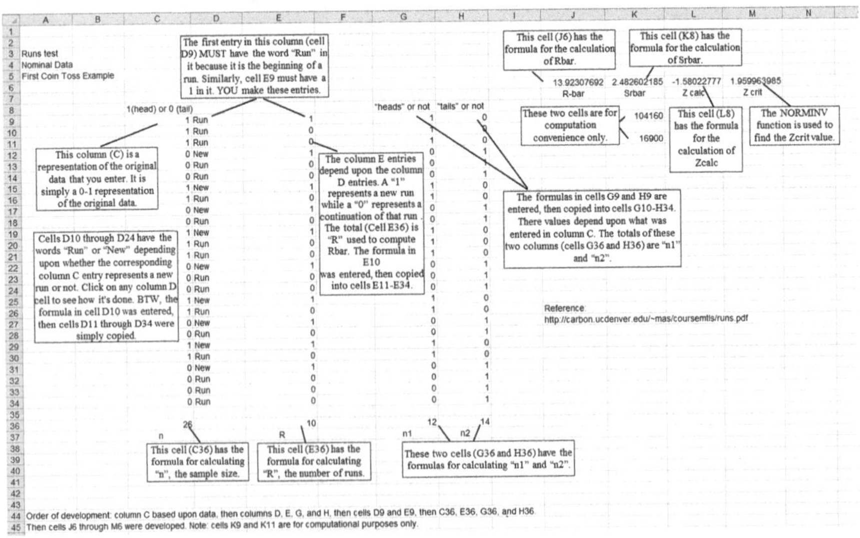

Figure 7.1: RUNS test spreadsheet—download ch7-ex.xls

You are (at least) 95% confident that you are making the correct decision regarding H_0, which states that the coin toss runs pattern observed was generate by a random process.

Notice that the Runs Test did NOT make the decision whether H_0 should be rejected or not. Accomplished in the spreadsheet, the Runs Test only provides information in an organized format so that the decision maker (you) can make a decision. And you can print the spreadsheet YOU develop if your boss wants to see the support for your decision.

Figure 7.1 shows the spreadsheet used to conduct the "Runs" test. This spreadsheet is part of "ch7-ex.xls" which can be downloaded for further inspection, as well as for copying.

The spreadsheet shown in Fig. 7.1 is difficult to read, so download it (ch7-ex.xls). Once downloaded, you can open it in OpenOffice, then click on individual cells to see how operations specific to the RUNS Test were accomplished. And you have a complete template to copy in support of YOUR conduct of future RUNS Tests.

This is the "beauty" of this concept. The book explains what the test accomplishes. The spreadsheet can be downloaded and you can click on the cells to see how the test was accomplished. It really can't get much simpler.

SIGN TEST

The Sign Test is used in a one population situation to determine whether a population median (usually of opinions) is equal to or not equal to a specific hypothesized number.

It makes very few assumptions about the nature of the distribution of the population being examined, so it has very general applicability. The sign test allocates a sign, either positive (+) or negative (−), to each observation according to whether it is greater or less than some hypothesized value (usually the hypothesized median value), and considers whether the number of "+"s and "−"s is substantially different from what we would expect by chance.

There are two separate cases of the Sign Test, depending upon sample size (n).

Any sample of size of less than 25 (but of at least 5) is considered to be small, and the Sign Test table in the Appendix can be used to determine the "critical" sign test value.

For a sample size of 25 or greater, the results of the Sign Test closely approximate the Normal probability distribution, so NORMINV can be used.

THAT'S ALL THERE IS TO IT!

EXAMPLE—Small Sample

Suppose that you are the manager of a fast food restaurant, and want to see whether your opinion that the median satisfaction score of customers who try (for free) a new product is correct or not. You think the median satisfaction score is 90. You want to be (at least) 95% confident that you are correct regarding your decision about your opinion, clearly stated in H_0, so α=0.05. You randomly select 12 people, have them try the new product, then offer a satisfaction score. The spreadsheet illustration in Fig. 7.2 shows the scores and procedure.

Notice that observed values equal to the hypothesized median, the ties, receive a 0 value for both differences, so they "go away."

The lesser of the differences totals (3) is NOT less than the "critical" value of 1 (α=0.05, N=11) found in the table, so H_0 CANNOT be rejected. There is not enough observed evidence from this sample to cause you to abandon your opinion that the satisfaction score median is 90.

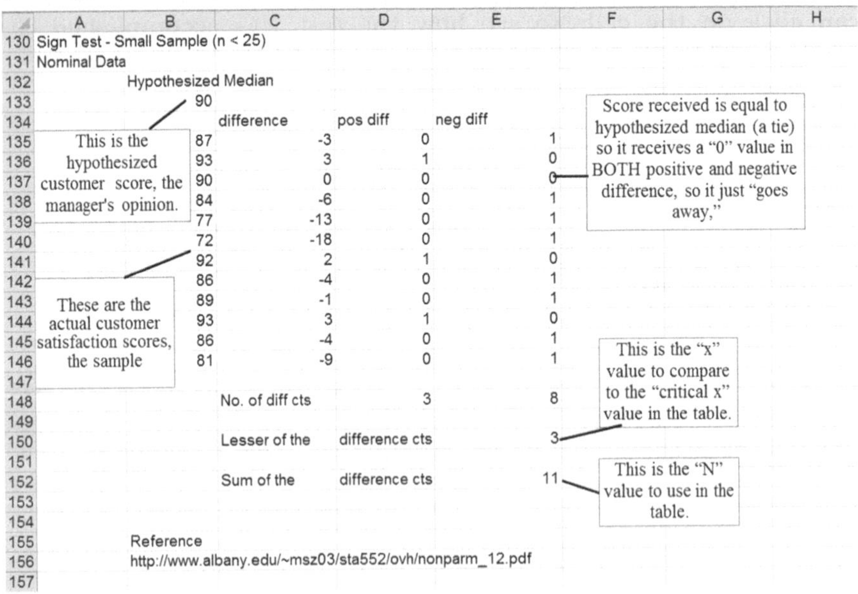

Figure 7.2: SIGN test—small sample—download ch7-ex.xls

ANOTHER EXAMPLE—Large Sample

For large sample (n $\geq$ 25), the calculated Sign Test value, Z_{calc}, is approximately normally distributed, so the NORMINV function can be used.

The formula for Z_{calc} is: ((X + 0.5) − (N − 2)) / (SQRT(N) / 2) where X and N are calculated by the Sign Test procedure: X = lesser of + or − count; N = sum of + and − counts. When counting, ties are ignored.

Note that "n" is the sample size, while "N" is the sum of + and − counts, ties ignored.

Again, this formula appears formidable. But don't worry, it can be copied.

If the (absolute value of the) calculated Z score is greater than the "critical" Z score returned by NORMINV, reject H_0. Otherwise, fail to reject H_0.

Say that as a manager of a large company, you think that your employees have a median score of 35 on some arbitrary classification score that does not connote any ranking. You randomly select 30 employees, classify them, and record the scores. You want to be (at least) 95% confident that you are correct regarding your decision about H_0, so α=0.05, and the NORMINV function is used to get Z_{crit} (so $\alpha/2$ is entered into the function). The spreadsheet illustrations in Fig. 7.3 shows the scores and procedure.

Again, observed values equal to the hypothesized median, the ties, receive a 0 value for both differences, so they "go away."

As can be seen, Z_{calc} of 7.52 is greater than Z_{crit} of 1.96, so H_0 CAN be rejected (note: you did not prove H_1 to be true) and you can be (at least) 95% confident that you are correct when you say the median classification score is NOT 35. What is the median classification score? The Sign Test cannot answer that question.

You CANNOT validly play "what if" by entering different median score values until H_0 cannot be rejected because analyzing the **SAME** sample numerous times inflates the probability of committing a Type I error. Solution: take another sample, change the median score, repeat the analysis.

ORDINAL DATA

WILCOXON SIGNED RANK TEST for ONE POPULATION

The Wilcoxon Signed Rank Test is used to make an inference about the median of one population. The procedure requires the specification of (what is thought to be) the median of the population of interest.

Sign Test - Large Sample (n => 25)						
Nominal Data						
	Hyp Score →	35				
			Difference	Pos Diff	Neg Diff	
	Actual	31	-4	0	1	
	Scores →	30	-5	0	1	
		22	-13	0	1	
		34	-1	0	1	
		15	-20	0	1	
		30	-5	0	1	
		23	-12	0	1	
		25	-10	0	1	
		35	0	0	0	
		15	-20	0	1	
		22	-13	0	1	
		26	-9	0	1	
		29	-6	0	1	
		24	-11	0	1	
		30	-5	0	1	
		25	-10	0	1	
		40	5	1	0	
		30	-5	0	1	
		19	-16	0	1	
		24	-11	0	1	
		30	-5	0	1	
		30	-5	0	1	
		37	2	1	0	
		27	-8	0	1	
		31	-4	0	1	
		21	-14	0	1	
		31	-4	0	1	
		28	-7	0	1	
		31	-4	0	1	
		30	-5	0	1	
			Difference Counts	2	27	
			Lesser Count	Difference	2	← x
				Sum of Differences	29	← N
				Z calc	7.527548262	
				Z crit	1.959963985	
	Reference					
	http://www.unm.edu/~marcusj/1Samplesign.pdf					

Figure 7.3: SIGN test—large sample—download ch7-ex.xls

EXAMPLE—Small Sample: n $\leq$ 20

Let's assume that you are the manager of a large company, and that you have ranked employees with respect to their effectiveness when dealing with potential customers, with higher rank scores indicating that they are more effective. Now you want to see if the median rank score assigned is 4. This is your null hypothesis: H_0 median score = 4. Alternately, the median rank score may not be 4: H_1 median score $\neq$ 4.

You randomly select nine (n = 9) employees, record their scores, and analyze the scores. The analysis is illustrated in Fig. 7.4.

The "rule" is that if the "W" value from the analysis is not less than the "critical W" value from the table, DO NOT reject H_0. If the "W" value from the analysis is LESS than the "critical W" value from the table, reject H_0.

IT'S THAT SIMPLE!

In the example below, the calculated "W" value is 15. This is the value compared to the "critical W" value from the appendix table, which (for n=9, α=0.05) is 5. Since the calculated "W" value of 15 is larger than (not less than) the "critical W" value of 5, do NOT reject H_0. There is no compelling evidence, based on the sample, that the median score $\neq$ 4. Make the managerial decision accordingly.

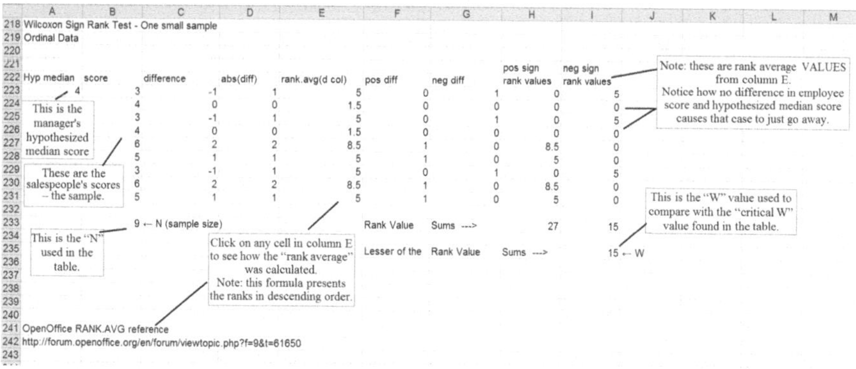

Figure 7.4: WILCOXON Sign Rank Test—small sample—download ch7-ex.xls

Use of this test really is simple and easy, especially since a spreadsheet you can download can be copied.

Two notes are in order here: (1) Samples of size 20 or smaller **MUST** use the procedure illustrated above and the table in the Appendix. Sample sizes less than 5 cannot be analyzed. Sample sizes greater than 20 can use either the table (if applicable) or the normal approximation method; (2) "W" is approximately normally distributed for samples of size greater than 20, so the NORMINV function can be used.

ANOTHER EXAMPLE—Large Sample: n $>$ 20

The situation described in the above example has, in this example, been expanded to 25 people's scores (n = 25) so that a large sample is created. The function NORMINV can be used for a normal approximation, as can the table. The spreadsheet is illustrated in Fig. 7.5.

Wilcoxon Sign Rank Test - one large sample
Normal Approximation; Ordinal Data
Hyp median 4

score	difference	abs(diff)	rank.avg(d column	pos diff	neg diff	pos sign rank values	neg sign rank values
4	0	0	4	0	0	0	0
5	1	1	13	1	0	13	0
4	0	0	4	0	0	0	0
6	2	2	20.5	1	0	20.5	0
3	-1	1	13	0	1	0	13
5	1	1	13	1	0	13	0
5	1	1	13	1	0	13	0
6	2	2	20.5	1	0	20.5	0
4	0	0	4	0	0	0	0
7	3	3	23.5	1	0	23.5	0
4	0	0	4	0	0	0	0
4	0	0	4	0	0	0	0
6	2	2	20.5	1	0	20.5	0
7	3	3	23.5	1	0	23.5	0
5	1	1	13	1	0	13	0
4	0	0	4	0	0	0	0
5	1	1	13	1	0	13	0
5	1	1	13	1	0	13	0
8	4	4	25	1	0	25	0
3	-1	1	13	0	1	0	13
5	1	1	13	1	0	13	0
4	0	0	4	0	0	0	0
3	-1	1	13	0	1	0	13
5	1	1	13	1	0	13	0
2	-2	2	20.5	0	1	0	20.5

25 ← N (sample size)

Rank Value Sums ---> 237.5 59.5

Lesser of the Rank Value Sums ---> 59.5 ← W

Z calc --> -2.43507551

Z crit → 1.95996398

Z calc numerator -90.5

denominator 37.1651719

FYI: Zcalc = W-(n*(n-1)/4) / sqrt(((n*(n+1)*(2n+1))/24))
Reference:
http://www.stat.ufl.edu/~ssaha/3024/CHAPTER14Examples.pdf

Figure 7.5: WILCOXON Sign Rank Test—large sample—download ch7-ex.xls

For a sample size of 25, a "W" value of 59.5 is calculated. From the table ($\alpha = 0.05$), the "critical" W value is 89, so, the H_0 CAN be rejected.

Also, as can be seen in the spreadsheet, the (absolute value of the) Z_{calc} value of -2.435 is greater than the Z_{crit} value of 1.959 returned by NORMINV. So, the H_0 CAN be rejected.

You CANNOT validly play "what if" by entering different median score values until H_0 cannot be rejected because analyzing the **SAME** sample numerous times inflates the probability of committing a Type I error. Solution: take another sample, change the median score, repeat the analysis.

NOTE: The Wilcoxon Signed Rank Test table goes as high as n = 30, but following these rules will "keep you out of trouble."

SUMMARY

You have been introduced to one population hypothesis testing using the RUNS Test, the SIGN Test, and the WILCOXON SIGN RANK Test.

You were introduced to:

- How to accomplish the Runs Test, the Sign Test, and the Wilcoxon Sign Rank Test
- Spreadsheets (that you can download) for all three tests
- How to interpret the various test results

EXERCISES

By the way, all the data presented here have already been keyed, and can be downloaded in "ch7-data.xls".

1. As the owner of a new car dealership, you have spent a lot of money on advertising to attract owners of car makes other than the ones you sell into your dealership. Is the advertising effective? To investigate the situation, you observe potential customers who drive into your dealership, and record whether or not he/she is driving a make you sell. You randomly select days and hours and record what make of car drive in (Y = make you sell, N = make you don't sell).
   The data are: Y N N N Y N N Y Y Y Y N N N Y N Y Y Y N N N N Y Y
   (a) State the null and alternative hypotheses.
   (b) Is the advertising effective or not? Show your analysis and how you reached your conclusion.
2. As the manager of a large retail store, you have spent a lot of money on advertising to address potential customer awareness of your store. You now say that the median awareness score is 100 (whatever that means). Is your opinion correct? In support of your opinion, you randomly select 20 customers as they are exiting some other store and ask for an overall awareness score.
   The data are: 90 110 105 85 95 100 105 120 150 85 100 110 90 75 105 120 100 105 90 100
   (a) State the null and alternative hypotheses.

(b) Is your opinion correct or not? Show your analysis and how you reached your conclusion.

3. As the manager of a large retail store, you have spent a lot of money on advertising to address potential customer shopping satisfaction. Higher satisfaction scores mean greater levels of satisfaction. You now say that the median satisfaction score is 10. In support of your opinion, you randomly select 20 customers as they are exiting and ask for an overall satisfaction score.
The data are: 9 8 11 7 10 11 9 10 12 7 5 11 14 9 10 11 10 12 8 9

 (a) State the null and alternative hypotheses.
 (b) Is your opinion correct or not? Show your analysis and how you reached your conclusion.

4. You are the local "outreach" director of a voter campaign. As such, before starting your "outreach" program, you want to assess whether the district that you serve has changed from its historical voting percentage of 75%. You randomly accost people, and ask them if they voted in the latest election (V = voted in latest election, S = skipped voting). Has the percentage changed?
The data are: S S V V V S V V V V V V S V V V V V S V V V V S V

 (a) State the null and alternative hypotheses.
 (b) Has the voting percentage changed or not? Show your analysis and how you reached your conclusion.

5. Let's use the same scenario as in No. 2, but this time you survey 50 customers.
The data are: 110 120 95 75 100 90 85 115 120 100 90 90 80 110 90 100 120 70 60 120 120 150 110 110 95 120 110 100 120 200 100 90 105 95 90 110 100 80 95 110 120 115 110 100 95 120 120 100 90 120

 (a) State the null and alternative hypotheses.
 (b) Is your opinion correct or not? Show your analysis and how you reached your conclusion.

6. Let's use the same scenario as in No. 3, but this time you survey 50 customers. The data are: 11 9 7 12 7 8 9 14 10 9 9 8 11 7 10 11 9 10 12 7 5 11 14 9 10 11 10 12 8 9 11 9 7 12 7 8 9 14 10 9 9 8 11 7 10 11 9 10 12 7

 (a) State the null and alternative hypotheses.
 (b) Is your opinion correct or not? Show your analysis and how you reached your conclusion.

8
Two Populations

There are two possible outcomes: if the result confirms the hypothesis, then you've made a measurement. If the result is contrary to the hypothesis, then you've made a discovery.
Enrico Fermi (1901–1954) Physicist

A statistician is a person whose lifetime ambition is to be wrong 5% of the time.

TWO POPULATION HYPOTHESIS TESTING

As with One Population Hypothesis Testing, this procedure focuses on making managerial decisions. All the concepts and rules you learned in Chap. 7 are still applicable. The 5-step hypothesis testing procedure is still applicable. The mutually exclusive and collectively exhaustive rules still apply, as do the assumptions of independence and randomness. And, as always, you need to establish your significance level BEFORE any analysis begins.

The Hypotheses:
H_0: $\mu 1 = \mu 2$ (the null hypothesis—both populations are equal with respect to what you are measuring)
H_1: $\mu 1 \neq \mu 2$ (the alternative hypothesis—the populations are not equal)

Another way to express the two hypotheses is:
H_0: $\mu 1 - \mu 2 = 0$ If you assume both populations to be equal, then subtracting one from the other will yield 0
H_1: $\mu 1 - \mu 2 \neq 0$

A primer on hypothesis testing can be found in the introduction.

W. Beatty, *Decision Support Using Nonparametric Statistics*,
SpringerBriefs in Statistics,
https://doi.org/10.1007/978-3-319-68264-8_8

INDEPENDENT and DEPENDENT SAMPLES

The terms "Independent Samples" and "Dependent Samples" are used throughout this chapter. Definitions are therefore warranted:

Independent Samples: These are samples drawn from two populations that have absolutely nothing in common except with respect to the factor being analyzed. For example, say you want to assess whether opinions are the same with respect to teaching methods: face-to-face versus via the Internet. If you sample from two different colleges (one teaching method from one college, the other teaching method from the other college), you get opinion scores from both groups with respect to teaching method. But, other factors, such as administration emphasis, peer pressures, and/or differences in people, will be included in the scores. There is absolutely nothing that can be done about this situation. The analysis includes all factors, so the analysis may or may not address the factor of interest.

There are two things that can be done to minimize this problem:

1. take large samples; and
2. take dependent samples

Dependent Samples: These are samples drawn from two populations that have in common the factor being analyzed, as well as other factor(s) that YOU define. For example, say you again want to assess whether opinions are the same with respect to teaching methods: face-to-face versus via the Internet. However, this time you ask the SAME people about their feelings regarding BOTH teaching methods (eliminating all people who have NOT utilized both teaching methods). This sampling method permits focus ONLY upon the factor being considered—opinions about teaching method. The samples are often called **"matched pairs"** samples. If, for example, you match the two teaching method score with respect to people, other factors, such as administration emphasis, peer pressures, and/or differences in people, are eliminated (or greatly reduced). Only the differences in teaching methodology opinions are analyzed. All other factor effects are minimized or eliminated.

"Before and After" rankings or scores are another example of dependent samples. For example, say you have people offer opinion scores for the face-to-face method. You then have the same people develop Internet courses, teach using the method, then offer opinion scores. A "before" (face-to-face) and "after" (Internet) situation is created. Matching the scores with respect to people eliminates (or minimizes) all factors except the one being analyzed.

Note that if the same people are asked for scores, but are not matched with respect to people, then only one of the three extraneous factors mentioned are eliminated: administration emphasis. The other two factors are still present: peer pressures, and/or differences in people. This is NOT an example of a dependent sample, or of matched pairs.

Why would independent samples EVER be drawn? Some situations dictate this sampling procedure. Independent samples sometimes simply cannot be avoided. It is ALWAYS better (when possible) to take dependent samples, as this method eliminates (or minimizes) extraneous factors.

Remember, YOU define whether samples (and therefore populations) or independent or dependent!

Consider this: If your two populations (and therefore your samples) are somehow related you can use the "MATCHED PAIRS" procedure. You define the relationship—the matching. For example, consider people "before and after" a taste test where "like" and "don't like" are the only two acceptable responses. Since you have the same people in both samples, you are eliminating (or at least greatly reducing) one source of variation—differences between people. With two independent populations (and therefore samples) you have variations in people AND variations caused by taking the taste test. With matched pairs, you have the same people both before and after the taste test, so the variation caused by people is eliminated or reduced. You can focus your attention entirely on the taste test.

NOMINAL DATA

INDEPENDENT SAMPLES

CHI-SQUARED 2×2 CONTINGENCY TABLE

Statistics Movies that Never Made it to the Big Screen: The Chi-Squared Syndrome

The Chi-Square (χ^2) Contingency Table can test differences between two independent samples to determine whether they were drawn from the same population or similar populations—H_0—with respect to only two response options. Hence the name 2×2. "2×2" means that the entire Chi-Squared Contingency Table is comprised of two rows and two columns: one for the two populations (say, for example, men and women, or children and adults), and one for the two response options (say taste or opinions). It really doesn't matter which is which.

The formula for χ^2 is very simple: Σ[(Observed – Expected)^2/Expected]

DEGREES OF FREEDOM (DF)

χ^2 also uses "degrees of freedom" (DF), which is simply: [(number of observed data columns – 1) * (number of observed data rows – 1)].

What is "Degrees of Freedom?"

First, consider only one row in a Chi-Squared Contingency Table. Some number of values (in the cells in the table) are free to vary (assume any value) for a given situation. For example, say that the total in a specific row (that has only two cells) is 100. One cell's value is free to assume any value between 0 and 100. The other cell's value is NOT free to assume any value if the total of 100 is to be maintained. Say one cell has a value of 65. The value in the other cell MUST be 35 in order for the total to be 100. That cell's value is NOT free to vary, hence one degree of freedom is "given up" for that row.

Next, consider the total for ALL rows in the table. All the row total values in all row are free to assume any values, except one whose value is determined (not free to assume any value) if the total for all rows is to be maintained. So the DF for rows is simply the total number of rows – 1 (one row total values not free to assume any value).

The reasoning for columns is identical. That reasoning holds regardless of the number of rows and columns in the Chi-Squared Contingency Table.

DF recognizes that the totals for all but one row and all but column are free to assume any value. Hence, the DF calculation formula is:

(number of data columns – 1) * (number of data rows – 1)

The name for this reasoning is "Degrees of Freedom", often written as DF.

DF is used in the OpenOffice CHIDIST function to calculate the probability of getting the calculated Chi-Square value given H_0 cannot be rejected. The CHIINV function also uses DF to return the exact p-value for a calculated Chi-Squared value.

CHI-SQUARED CONTINGENCY TABLE EXAMPLE

Suppose you are the manager of a company that has developed a new soft drink, and wonder if men and women are similar with regard to liking it or not. Your opinion is that men and women are similar regarding liking the new drink: H_0: $\mu_{\text{men}} = \mu_{\text{women}}$. You randomly select 40 men and 90 women for your test, as illustrated below:

	Men	Women
Like	10	50
Don't Like	30	40

The spreadsheet illustrated in Fig. 8.1 has three sections: Observed Data, Expected Data, and Chi-Square Calculations.

	A	B	C	D	E	F	G	H	I	J	K	L
1												
2	Chi-Squared Test, Two Independent Populations											
3	Nominal Data											
4												
5		Men	Women									
6		Observed Data		Totals		Expected Data		Totals		Chi-Square Calculation		Totals
7	Like	10	50	60		18.46153846	41.53846154	60		3.878205128	1.723646724	5.601851852
8	Don't Like	30	40	70		21.53846154	48.46153846	70		3.324175824	1.477411477	4.801587302
9	Totals	40	90	130		40	90	130				
10												
11								This is the calculated Chi-Sq value, the test statistic: →				10.40343915
12								It is the sum of cells L7 and L8, which are the sums of				
13								cells J7&K7 and J8&K8				
14												
15						This is the prob of getting the calculated Chi-Sq value given Ho cannot be rejected: →						0.001257808
16						This is a "P-value" returned by the CHIDIST function.						
17												

Figure 8.1: χ^2 test of independence—download ch8-ex.xls

- The Observed Data section simply records the sample results. The 'totals' row and column are ONLY for Expected Data calculations, as you will see.
- The Expected Data section calculates what results can be expected if H_0 should not be rejected. For each value (cell) it is simply: [(corresponding row total * corresponding column total)/entire sample size] the totals for this section should match the previous totals. If not, something is wrong.
- The Chi-Square Calculations section calculates the χ^2 value for individual cells.

The individual cell χ^2 values are then summed to create the χ^2 value for the entire table. Download the Chap. 8 examples (ch8-ex.xls), then explore to see how this was accomplished.

As can be seen in the 2×2 (2 categories—men and women; 2 response possibilities) spreadsheet, the calculated *p*-value (0.001257808, returned using CHIDIST) for the calculated χ^2 value (10.40344) is below α=0.05, so you can (with at least 95% confidence that you are correct) reject H_0. Observed data inspection (permitted because H_0 was rejected) shows that men and women are dissimilar with respect to liking the new soft drink. Women like the new soft drink quite a bit more than men, while men and women are similar with respect to not liking it.

Note that the *p*-value of 0.001257808, or 0.1258% is consistent with the table in the Appendix. The "critical" χ^2 value (DF=1, α=0.001) from the table is 10.83. The calculated χ^2 value is 10.4, and has a *p*-value of 0.001258, just slightly greater than 0.001. The "critical" χ^2 value has exactly a 0.1% probability of occurrence (α = 0.001, *p*-value = 0.001), and the calculated χ^2 value has just over a 0.1% chance of occurrence if H_0 cannot be rejected.

Strictly speaking, H_0 cannot be rejected with 99.9% confidence if using the table. But consider this. A table value as great as 10.83 has exactly a 0.1% chance of occurrence if H_0 cannot be rejected. The calculated χ^2 value of 10.40343915 has a 0.1257808% chance of occurrence if H_0 cannot be rejected. The point (and the only point) here is that the two methods of hypothesis testing yield (virtually) identical results—10.4 is not much less than 10.83, and 0.125% is not much greater than 0.1%.

Isn't that comforting? It should be. The two methods yield (virtually) identical results. "But wait," you protest. Decision-making requires unambiguous data analysis. Now the ambiguous word "virtually" is used. What gives? The point here is that you can arrive at very, very close conclusions using either method since they are consistent. And the decision about H_0 would have been identical had α been 0.05 or 0.01.

So, which way is best for conducting an hypothesis test? The answer is "Yes!" The best way is the one with which you are most comfortable. Using either the table or the CHIDIST function yields consistent results. That is the only point of this entire verbose discussion.

Note that Chi-Square ONLY tests whether men and women are similar with regard to liking the new drink—nothing more. But, because H_0 was rejected, data inspection is permitted. It appears that women like the new soft drink more than men. Make your managerial decision accordingly. For example, you may want to restrict advertising to only women's magazines.

CHI-SQUARED 2×n CONTINGENCY TABLE

This method works for χ^2 tables of ANY size.

Now extent the above scenario so that there are five possible taste test responses: Great Dislike, Dislike, Neutral, Like, Great Like. (Yes, the taste test responses constitute a Likert scale, considered to be ordinal data. But ordinal data can be analyzed using a nominal data technique.) With these preference possibilities, are men and women similar? The χ^2 2×5 (2 categories—men and women; 5 response possibilities) spreadsheet is illustrated in Fig. 8.2.

The degrees of freedom calculation applies for the χ^2 2×n Contingency Table as well. The DF for the above table is 4: ([2 columns – 1] * [5 rows – 1])

The spreadsheet again has three sections: Observed Data, Expected Data, and Chi-Square Calculation. The calculated χ^2 value is 27.347...., and its

Chi-Square Test, Two Independent Populations
Nominal Data

	Men	Women							
	Observed Data		Totals	Expected Data		Totals	Chi-Square Calculation		Totals
Great Dis	10	50	60	22.10526316	37.89473684	60	6.629072682	3.866959064	10.49603175
Dislike	30	40	70	25.78947368	44.21052632	70	0.687432868	0.401002506	1.088435374
Neutral	40	30	70	25.78947368	44.21052632	70	7.830290011	4.567669173	12.39795918
Like	20	40	60	22.10526316	37.89473684	60	0.200501253	0.116959064	0.317460317
Great Like	5	20	25	9.210526316	15.78947368	25	1.92481203	1.122807018	3.047619048
Totals	105	180	285	105	180	285			

This is the calculated Chi-Sq value, the test statistic: → 27.34750567

This is the prob of getting the calculated Chi-Sq value given Ho cannot be rejected: → 1.69085E-05

Figure 8.2: χ^2 2×5 contingency table—download ch8-ex.xls

probability of occurrence if H_0 can be rejected is 0.00169%. This is everything you need to make a decision.

The χ^2 Contingency Table can be ANY size, from 2×2 to 2×n, where 2 is the number of categories being examined (such as men and women, or yes and no, or tall and short, etc.), and "n" is the number of possible responses or characteristics. If H_0 is rejected, problems arise, however, with observed data inspection and interpretation if "n" is much greater than four. For example, in the spreadsheet shown in Fig. 8.2, H_0 can be rejected (thus permitting data interpretation), but, as you can see, interpretation is difficult.

WARNING

χ^2 Contingency Table analysis results should NOT be interpreted if ANY observed data have a count of less than five in any one (or more) category (or cell). So, if there is a value less than 5 in any cell(s) in the entire table, what can be done? Say, for example, there are only three women who respond with "Great Dislike." This will invalidate χ^2 analysis. There are two possible alleviations:

1. Continue to sample until the problem disappears. The counts in other categories may increase as well, but that situation is acceptable.

2. If further sampling is not possible (like for time or cost reasons), then "collapsing" is the solution. For example, say that 15 women responded "Dislike" and only 3 women responded "Great Dislike." Collapsing the two categories—"Great Dislike" and "Dislike"—into one category (call it "Unfavorable" with counts 3 + 15 = 18) will alleviate the problem. Analysis can then proceed. The collapsing must also be done for the men as well, even if no problem existed. Collapsing must be accomplished consistently in ALL rows and/or columns.

CHI-SQUARE CORRECTION

A general note about χ^2 analysis is appropriate here. The χ^2 computation can be "corrected" to decrease the likelihood of error in estimating probabilities when small expected frequencies are calculated. What is small? Some statisticians say if the expected frequency is less than 5. When should the correction be applied? Some statisticians say never, some say always, some say only when ANY expected frequency calculation is small. The correction is simply the subtraction of 0.5 from the absolute value of what is observed from what is expected before that quantity is squared. It is mentioned here only to make you aware of its existence.

NONE of the χ^2 Contingency Tables illustrated here are corrected.

2×5 EXAMPLE

Let's assume that you like to play BINGO, and that you frequent one particular "BINGO Parlor." Recent observations have led you to question the randomness with which numbers are called. You, therefore, count the number of times a letter is called alone and the number of times that letter is called more than once. You record the letters being called in randomly selected games. The results are:

N I N B B G O G O N B O I I I B N N G G G G N B G G B O
I B I I B G O O O G N B I I O O N B G N G O B G N B B B G
O N G O N I B B I G B B I O O G O B B G G G O O O N N G B
B G O O N N N B O G G N N N B B O O G O B I I I I N G N N
N G G B I O O I I O I I N G I G I O B G G O B G I I I O I I G G O
O I I I B G G N N N I I G G G O B G O O B B I N N N G G B O G I G G N I

This situation is quite similar to the Runs Test from Chap. 7, but the Runs Test will not work here because there are two populations (repeated letter and not repeated letter), and five outcome possibilities: B, I, N, G, and O (Fig. 8.3).

H_0: the Bingo game is fair, the letters were called in a random pattern
H_1: the Bingo game is NOT fair, the letters were NOT called in a random

Chi-Square Test, Two Independent Populations									
Nominal Data									
	Repeated Letter Called?								
	Yes	No							
	Observed Data		Totals	Expected Data		Totals	Chi-Square Calculation		Totals
B	8	15	23	9.663865546	13.33613445	23	0.286474242	0.20759003	0.494064272
I	9	11	20	8.403361345	11.59663866	20	0.042361345	0.030696626	0.073057971
N	9	10	19	7.983193277	11.01680672	19	0.129509067	0.09384715	0.223356217
G	11	18	29	12.18487395	16.81512605	29	0.115218777	0.083491868	0.198710645
O	13	15	28	11.76470588	16.23529412	28	0.129705882	0.09398977	0.223695652
Totals	50	69	119	50	69	119			
							This is the calculated Chi-Sq value, the test statistic: →		1.212884757
				This is the prob of getting the calculated Chi-Sq value given Ho cannot be rejected: →					0.875972705

Figure 8.3: χ^2 BINGO example—download ch8-ex.xls

pattern
$\alpha = 0.05$

The p-value of 0.87597, or 87.597% (for the calculated χ^2 value of 1.21288) is greater than 5% (α=0.05), so H_0 CANNOT be rejected. You conclude with at least 95% confidence, that, from this sample, the Bingo game is fair.

MATCHED PAIRS or DEPENDENT SAMPLES

McNEMAR TEST

Developed by Quinn McNemar and presented in 1947, this test yields reliable results ONLY for matched pairs samples for which only two outcomes are possible. Therefore the McNemar test considers only 2×2 data. Although the data presented in tabular form looks similar to a χ^2 2×2 contingency table, the two tests are not equivalent.

The McNemar does not test for independence as does χ^2, but looks for consistency in responses regarding two variables. It determines if a statistically significant change has occurred on a dichotomous variable, the matching variable.

To illustrate what it does, consider this business training situation. All salespeople (the matching variable) receive some training so they can better achieve customer satisfaction. Those who achieve 'high scores' in customer satisfaction receive additional training (the factor being investigated).

But the additional training is expensive, so you want to see if it is effective. To see if additional training results in higher numbers of satisfied customers, you randomly select 165 salespeople from ALL of them and examine them, 91 of them have received additional training and 74 have not. For the 165 salespeople, you ask customers who were served by them if they provided satisfactory sales service or not. The results are illustrated below:

	Sat Svc	Unsat Svc
No Addl Tng	50 (a)	24 (b)
Addl Tng	86 (c)	5 (d)

In the above table, the salespeople not additionally trained are categorized in the first row, of whom 50 gave satisfactory service (a) and 24 did not (b). Those additionally trained are categorized in the second row, of whom 5 did not provide satisfactory service (d) and 86 did (c).

The results are illustrated in Fig. 8.4.

	A	B	C	D	E	F	G
287	McNemar Examples						
288							
289	Original Example						
290		sat svc	not sat svc	totals			
291	No Addl Tng	50	24	74			
292	Addl Tng	86	5	91			
293				165			
294	test stat value	33.82727273	← compare this value to the “critical” X2 table value, DF=1				
295			Click on cell B295 to see how the test stat value was calculated				
296							
297							
298	Original Example stated another way						
299		not sat svc	sat svc	totals			
300	Addl Tng	5	86	91			
301	No Addl Tng	24	50	74			
302				165			
303	test stat value	33.82727273	← compare this value to the “critical” X2 table value, DF=1				

Figure 8.4: First McNemar example—download ch8-ex.xls

The McNemar test recognizes that some salespeople will move, regardless of having been additionally trained or not, from the Satisfactory to Not Satisfactory category, while other salespeople will move from Not Satisfactory to Satisfactory category randomly when the customers rate them on providing satisfactory service or not. If the training is NOT effective, the number of salespeople who move from the Not Satisfied to the Satisfied category should be about equal to those who move from the Satisfied to the Not Satisfied category. But if the additional training is effective, if there is a direction to the movement, the McNemar test will observe it.

Actually, the 50 salespeople in the ‘No Addl Tng - sat svc’ category (a) and the 5 salespeople in the ‘Addl Tng - not sat svc’ category (d) are irrelevant to this test.

The null hypotheses is H_0: $\mu 1 = \mu 2$, that is, the random movement is the SAME both ways, therefore the additional training is NOT effective.

The alternative hypotheses is H_1: $\mu 1 \neq \mu 2$, that is, the random movement is NOT the same both ways, therefore the additional training IS effective.

The test statistic χ^2 is calculated as: (absolute value of $(\mathrm{b}-\mathrm{c})-1)^2)/(\mathrm{b}+\mathrm{c})$ with the 'b' and 'c' values are identified from the following table:

	After-present	After-absent
Before-present	a	b
Before-absent	c	d

Table cells 'b' and 'c' are called the discordant cells because they represent pairs with a difference and reflect the change (or absence of it) due to the factor being investigated. Cells 'a' and 'd' are the concordant cells. These cells do not contribute any information about a difference between pairs or over time so they aren't used to calculate the test statistic. Their values can actually be omitted, but are included here to better illustrate what's going on.

This is the KEY to understanding the McNemar test. How the contingency table is represented doesn't matter so long as the **CHANGE** due to the factor being investigated (like additional training) appears in cells 'b' and 'c'.

The test statistic is distributed as χ^2 with ONE degree of freedom.

In the example above, for salespeople who had NOT received additional training (the 'after' group in this example), 50 (a) of them gave satisfactory service (present), while 24 (b) did not (absent). Of those that did receive additional training (the 'before' group), 86 (c) of them gave unsatisfactory service (absent), while 5 (d) did (present).

Look at the spreadsheet illustration in Fig. 8.4. The first part represents the example as stated above. The calculated test statistic χ^2 is 33.827.... The "critical" values of χ^2 for 1 degree of freedom is, from the table, 10.83 for α=0.001. Since the value of the calculated test statistic exceeds the critical value for α=0.001, you can reject H_0 and conclude with at least 99.99% confidence that the additional training is effective in increasing satisfied customers with respect to sales service.

The second part solves the example problem, but with a different contingency table. Notice that the discordant cells (b and c) still represent the values "caused" by the factor being investigated. The test statistic value is the same, so an understanding of how the test works is mandatory.

BEFORE and AFTER

Now a "Before and After" example.

Note: this "before and after" is concerned with the matched pairs of the populations and are NOT the same as in the McNemar test.

Let's say we randomly select 155 salespeople and ascertain that 108 of them provide satisfactory service and 47 don't. These salespeople undergo intensive sales training. After training, we ascertain that 120 of these same salespeople (hence matched pairs) provide satisfactory service and 35 don't.

H_0: $\mu 1 = \mu 2$, or the before and after performances are the same, and that the training is ineffective
H_1: $\mu 1 \neq \mu 2$, or the before and after performances are NOT the same, and that the training is effective α is whatever you are comfortable with

The calculated test statistic value of 1.4756... is not greater than the χ^2 "critical" value (for α=0.05) of 3.84, so H_0 CANNOT be rejected. You can be at least 95% confident that you are making the correct decision regarding H_0. Because H_0 was not rejected, you must conclude that there is no difference between before and after performance and that the training is not effective.

The spreadsheet is illustrated in Fig. 8.5.

Before and After				
	sat svc		not sat svc	
Before Tng		108	47	155
After Tng		120	35	155
test stat value	31.04191617 ← compare this value to the "critical" X2 table value, DF=1			

Figure 8.5: McNemar 'before and after'—download ch8-ex.xls

PROPORTIONS

Now a proportions example. Let's say that we want to study training company-wide, and that thousands of salespeople are examined before and after training, just like the situation described above. Let's say that 62% of them provide satisfactory service. Let's say that 81% of them provide satisfactory service after training. Based on this data, can you conclude that company-wide training has been successful?

The spreadsheet is illustrated in Fig. 8.6.

Proportions		
	sat svc	not sat svc
Before	62	38
After	81	19
test stat value	14.82352941 ← compare this value to the "critical" X2 table value, DF=1	

Figure 8.6: McNemar proportions—download ch8-ex.xls

The test statistic value of 14.823... exceeds the 0.01 "critical" value of 6.64, so you are 99% confident that you are correct when you reject H_0. You are more than 95% confident, but less than 99% confident that the training is effective.

SIGN TEST

The Sign Test is used in a two population situation to determine whether there is a real difference between matched pairs of data, such as "before and after" measurements. It makes very few assumptions about the nature of the distribution of the populations being examined, so it has very general applicability. The Sign Test allocates a sign, either positive (+) or negative (−), to each pair of observation according to whether they are greater or less than some hypothesized (expected) change in value (such as an increase), and considers whether the observed signs are substantially different from what we would expect by chance.

EXAMPLE—SMALL SAMPLE: sample size less than 25

Sign Test - Two Populations
Nominal Data
Small Sample - Sample Size < 25

Before Tng	After Tng	Difference	Pos Diff	Neg Diff	
6	7	1	1	0	
5	7	2	1	0	
6	6	0	0	0	Notice how, when the before and after scores are the same, the difference just "goes away"
5	4	-1	0	1	
4	5	1	1	0	
5	8	3	1	0	
6	7	1	1	0	
5	6	1	1	0	
5	7	2	1	0	
3	3	0	0	0	
4	3	-1	0	1	
4	5	1	1	0	
5	6	1	1	0	
6	7	1	1	0	
		No. of → Differences	10	2	
		Lesser of → Differences		2 ← "x"	← This is the value to compare to the "critical" table value
		Sample Size		14 ← "N"	← This is the "N" value to use

Figure 8.7: SIGN test—small sample—download ch8-ex.xls

Suppose that you are the manager of a large company that is conducting "sensitivity training." You randomly select 14 employees, evaluate them and assign a "before training sensitivity score," train them, then evaluate them and assign an "after training sensitivity score." Now you want to see if the training is effective, that "sensitivity" has increased. Your initial opinion is that the training does NOT increase "sensitivity"—H_0: $\mu_{before} = \mu_{after}$. The data are illustrated in the spreadsheet shown in Fig. 8.7.

The Sign Test appendix table is based upon the scores observed in the sample observed. If the test statistic value (x) is greater than the table value for α, then H_0 cannot be rejected.

In the spreadsheet shown in Fig. 8.7, you can see that the calculated test statistic χ^2 is EQUAL TO the table value (n=14, α=0.05) of 2, so the H_0 CAN be rejected. You can be (at least) 95% (α=0.05) confident that you are making the correct decision regarding H_0. Based upon this sample, your initial H_0: training is NOT effective, can be rejected. Make your managerial decision accordingly.

EXAMPLE—LARGE SAMPLE: sample size $\geq$ 25

If the sample is large (N $\geq$ 25) then the calculated Sign Test statistic is approximately normally distributed, so NORMINV can be used. The Z_{calc} test statistic formula is:

Z_{calc} is: $((x + 0.5) - (N - 2)) / (\sqrt{N} / 2)$

where x and N are calculated by the Sign Test procedure: x= lesser of (+) or (−) count; N = sum of (+) and (−) counts, ties ignored. (just like in Chap. 7)

Assume the same "sensitivity training" scenario as above, but this time you randomly select 30 employees, evaluate them and assign a "before training sensitivity score," train them, then evaluate them and assign an "after training sensitivity score." Now you want to see if the training is effective, that "sensitivity" has increased. Your initial opinion is that the training does NOT increase "sensitivity": H_0: $\mu_{before} = \mu_{after}$. The data are illustrated in the spreadsheet shown in Fig. 8.8.

From the spreadsheet shown in Fig. 8.8 you can see that (the absolute value of) Z_{calc} of −8.21 is greater than Z_{crit} of 1.96 (using NORMINV), so based upon this sample, your initial H_0: training is NOT effective, can be rejected. Make your managerial decision accordingly.

Sign Test - Two Populations						
Nominal Data						
Large Sample - Sample Size > 25						
Before Tng	After Tng		Difference	Pos Diff	Neg Diff	
6	7		1	1	0	
5	7		2	1	0	
6	6		0	0	0	
5	4		-1	0	1	
4	5		1	1	0	
5	8		3	1	0	
6	7		1	1	0	
5	5		0	0	0	
5	7		2	1	0	
3	3		0	0	0	
4	3		-1	0	1	
4	5		1	1	0	
5	6		1	1	0	
6	7		1	1	0	
4	4		0	0	0	
7	6		-1	0	1	
2	5		3	1	0	
2	3		1	1	0	
5	8		3	1	0	
7	5		-2	0	1	
3	4		1	1	0	
5	5		0	0	0	
6	5		-1	0	1	
4	5		1	1	0	
7	8		1	1	0	
2	6		4	1	0	
5	5		0	0	0	
5	6		1	1	0	
3	4		1	1	0	
5	6		1	1	0	
			No. of → Differences	19	5	
			Lesser of → Differences		5	← "x"
			Sample Size		30	← "N"
			Z calc →		-8.215838363	
			Z crit →		1.959963985	

Figure 8.8: SIGN test—large sample—download ch8-ex.xls

ORDINAL DATA

INDEPENDENT SAMPLES

MANN-WHITNEY U TEST

MANN-WHITNEY "U" TEST with "small" samples

The Mann-Whitney U Test analyzes independent, unmatched samples. It is used to test the null hypothesis that two samples come from the same population or similar populations.

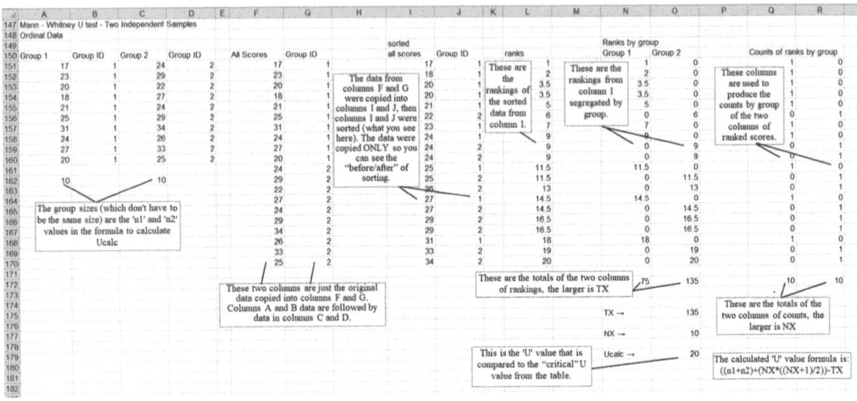

Figure 8.9: MANN-WHITNEY "U" test—small sample—download ch8-ex.xls

Let's assume that you are the sales manager for a large company. You randomly select 10 sales employees from the east coast office (group 1) and 10 sales employees from the west coast office (group 2). (Note: groups do NOT have to be the same size.) Each employee is scored on his/her sales effectiveness, with higher scores indicating more effective. Do east coast salespeople differ from west coast salespeople with respect to sales effectiveness? The H_0 is: east coast and west coast salespeople are equally effective.

The spreadsheet to calculate U is illustrated in Fig. 8.9.

U_{calc} must be EQUAL to or LESS than this critical value (U_{crit}) found in the table in the Appendix in order to reject H_0. U_{calc} in the spreadsheet shown in Fig. 8.9 is 20.0. U_{crit} from the appendix table (for n1=10 and n2=10, α=0.05) is 23, so H_0 CAN be rejected. East coast and west coast salespeople are not the same in terms of effectiveness. Note that the amount of the difference is NOT given by M-W.

SORTING

The MANN-WHITNEY Test requires the SORTING of data, so here is a tutorial.

The spreadsheet shown in Fig. 8.9 appears complicated, but it is really rather simple. The only "tricky" part is sorting, but even that is easy once accomplished. Here's how!

Start with data and group (or sample) identifiers. The process is illustrated in Figs. 8.10, 8.11, 8.12, 8.13, and 8.14.

Step 1—copy the original data and group identifiers into continuous columns.

Step 2—copy THAT data into contiguous columns so it can be sorted.

Step 3—highlight (by clicking and dragging) the data to be sorted (so the Sort procedure will know what to sort—in this example, columns I and J).

Step 4—click on "Data," then "Sort" to open the 'Sort Criteria' window in OpenOffice

Step 5—First, click on the 'Sort By' down arrow. Second, click on Column I (we want to sort based on column I in this example) to select it. Third, click on "OK"

After step 5 (Fig. 8.14), the data are sorted and are ready for subsequent use. (Note 1—the sorted data remain in columns I and J) (Note 2—Step 2 can eliminated once familiarity with Sort is achieved)

Once sorted (columns I and J), the data are ranked (column L) exactly as with the WILCOXON Sign Rank Test above. The rankings are then segregated into ranks by group (columns N and O), then totaled (cells N173 and

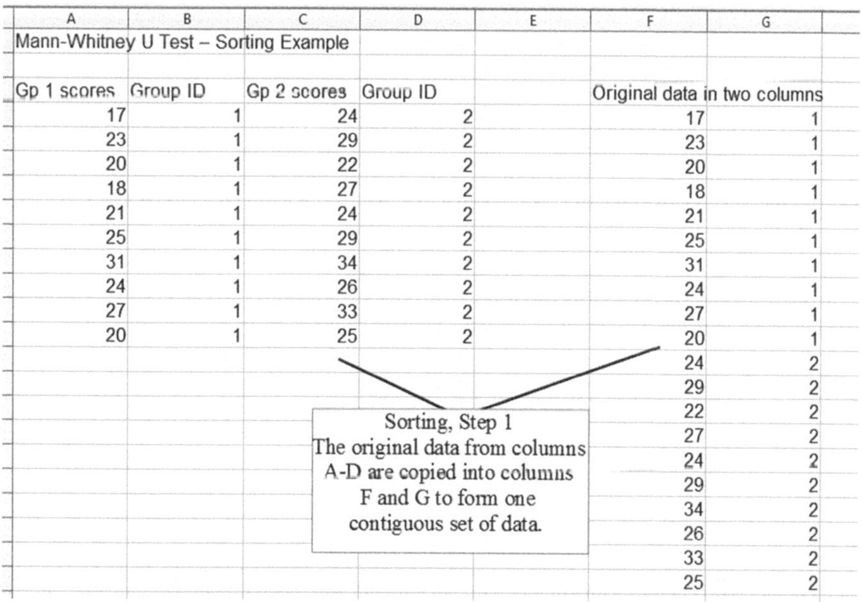

A	B	C	D	E	F	G
Mann-Whitney U Test – Sorting Example						
Gp 1 scores	Group ID	Gp 2 scores	Group ID		Original data in two columns	
17	1	24	2		17	1
23	1	29	2		23	1
20	1	22	2		20	1
18	1	27	2		18	1
21	1	24	2		21	1
25	1	29	2		25	1
31	1	34	2		31	1
24	1	26	2		24	1
27	1	33	2		27	1
20	1	25	2		20	1
					24	2
					29	2
					22	2
					27	2
					24	2
					29	2
					34	2
					26	2
					33	2
					25	2

Figure 8.10: MANN-WHITNEY "U" test—sorting, step 1—download ch8-ex.xls

F	G	H	I	J
Original data in two columns			Original data in two columns	
17	1		17	1
23	1		18	1
20	1		20	1
18	1		20	1
21	1		21	1
25	1		22	2
31	1		23	1
24	1		24	1
27	1		24	2
20	1		24	2
24	2		25	1
29	2		25	2
22	2		26	2
27	2		27	1
24	2		27	2
29	2		29	2
34	2		29	2
26	2		31	1
33	2		33	2
25	2		34	2

Sorting, Step 2
This step copies the data from columns F and G into columns I and J ONLY to provide a 'pre-post' data view. Columns I and J will actually be sorted.

Figure 8.11: MANN-WHITNEY "U" test—sorting, step 2—download ch8-ex.xls

O173 in this example—the larger total is TX in the U_{calc} formula below). The ranks are counted for each group (columns Q and R—these totals ['N1' and 'N2'], the larger of which is NX in the U_{calc} formula below).

Finally, U_{calc} (cell O180 in this example) is calculated. This is the value to compare to the "critical" value in the table. The formula to calculate U_{calc} is:

$$U_{calc} = ((n1*n2)+(NX*((NX+1)/2))-TX$$

where 'n1' and 'n2' are the counts of the two group, NX and TX are calculated in the spreadsheet.

WOW! All of this sounds complicated. But it's really not. First, download the spreadsheet in which this is located (ch8-ex.xls) to see how all of this was accomplished.

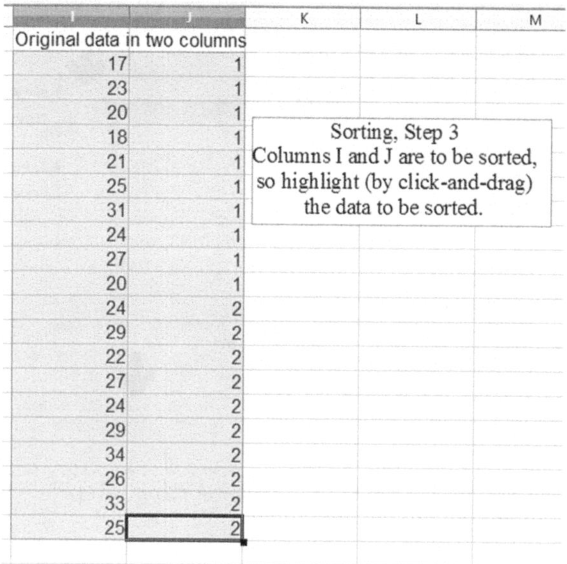

Figure 8.12: MANN-WHITNEY “U” test—sorting, step 3—download ch8-ex.xls

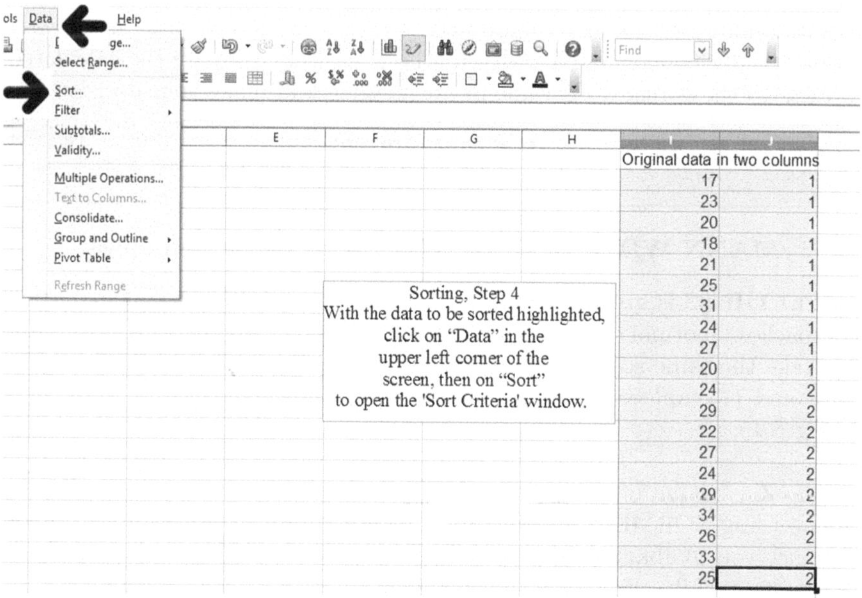

Figure 8.13: MANN-WHITNEY “U” test—sorting, step 4—download ch8-ex.xls

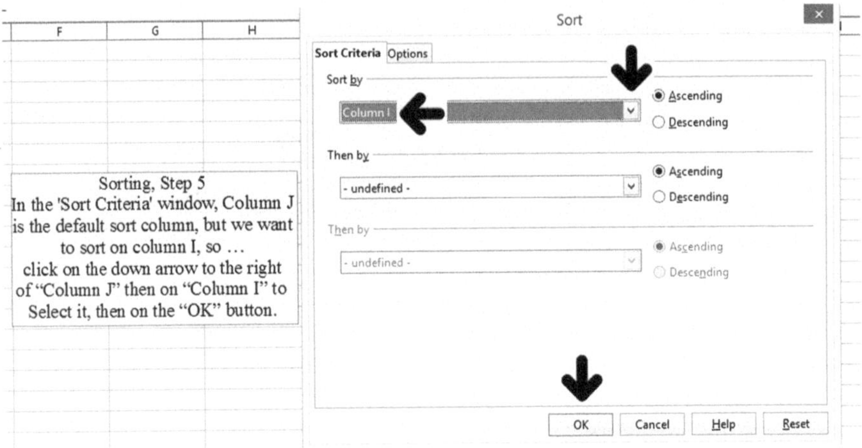

Figure 8.14: MANN-WHITNEY "U" test—sorting, step 5—download ch8-ex.xls

Then, **to conduct your own Mann-Whitney test**, all you have to do is put your samples data into columns A and C, sort it, then copy cells L152 through R152 to fit your sorted data. Next, copy cells M173 through R180 into your spreadsheet and adjust them so that they reference your data.

The ranking totals and ranking counts, as well as TX, NX, and U will automatically be calculated. All you'll have to do is compare the calculated U value to the "critical" U value from the table and make your decision.

It's that simple!

MANN-WHITNEY "U" TEST with "large" samples

If n1 **OR** n2 is greater than 10, then the Mann-Whitney U distribution approaches a normal distribution, and NORMINV can be used. As an example, take the same scenario as above, but you select 20 employees from one coast and 17 employees from the other coast. The spreadsheet is illustrated in Fig. 8.15.

The Z_{calc} value of 1.35618... is NOT larger than the Z_{crit} value of 1.96 (found using NORMINV), so you CANNOT reject H_0. These samples do not convince you to abandon your initial assumption that salespeople from both coasts are equally effective with customers.

Notice that this result is consistent with the result reached if the appendix table is used. Why would the approximation method ever be used? If n1 and/or n2 exceed the table sample sizes, this method must be used.

Notice also that the ONLY addition to the first spreadsheet illustrated was the calculations of Z_{calc} and Z_{crit} values. So using this method is "no big deal."

Sample (Group) sizes n1 and n2 do NOT have to be the same size in order to use either method.

MATCHED PAIRS or DEPENDENT SAMPLES

WILCOXON SIGN RANK TEST

For the difference between two populations in a MATCHED PAIRS (before and after) setting, use the WILCOXON SIGNED RANK TEST for ONE POPULATION (the differences) that was illustrated in Chap. 7. If the number of differences ("n") is 30 or less, the table can be used, OR the normal approximation (shown below) can be used. If "n" is greater than 30, the normal approximation must be used. The spreadsheet shown in Fig. 8.16 illustrates the procedure.

Note that there are TWO populations represented here, and that the scores are matched. But, when differences are considered, **the differences can be analyzed as if they represent a single population, so the**

Mann - Whitney U test - Two Independent Samples
Ordinal Data
Normal Approximation

Group 1	Group ID	Group 2	Group ID	All Scores	Group ID	sorted all scores	Group ID	ranks	Ranks by group Group 1	Group 2	Counts of ranks by group	
16	1	20	2	16	1	16	1	1.5	1.5	0	1	0
20	1	17	2	20	1	16	2	1.5	0	1.5	0	1
20	1	22	2	20	1	17	2	4.5	0	4.5	0	1
18	1	20	2	18	1	17	2	4.5	0	4.5	0	1
21	1	21	2	21	1	17	2	4.5	0	4.5	0	1
24	1	23	2	24	1	17	2	4.5	0	4.5	0	1
31	1	31	2	31	1	18	1	7	7	0	1	0
23	1	26	2	23	1	19	1	8	8	0	1	0
27	1	16	2	27	1	20	1	12	12	0	1	0
20	1	17	2	20	1	20	1	12	12	0	1	0
21	1	21	2	21	1	20	1	12	12	0	1	0
30	1	17	2	30	1	20	1	12	12	0	1	0
19	1	20	2	19	1	20	2	12	0	12	0	1
26	1	24	2	26	1	20	2	12	0	12	0	1
29	1	17	2	29	1	20	2	12	0	12	0	1
24	1	21	2	24	1	21	1	17.5	17.5	0	1	0
25	1	24	2	25	1	21	1	17.5	17.5	0	1	0
30	1			30	1	21	2	17.5	0	17.5	0	1
20	1			20	1	21	2	17.5	0	17.5	0	1
22	1			22	1	22	1	20.5	20.5	0	1	0
				20	2	22	2	20.5	0	20.5	0	1
				17	2	23	1	22.5	22.5	0	1	0
				22	2	23	2	22.5	0	22.5	0	1
				20	2	24	1	26	26	0	1	0
				24	2	24	1	26	26	0	1	0
				23	2	24	2	26	0	26	0	1
				31	2	24	2	26	0	26	0	1
				26	2	24	2	26	0	26	0	1
				16	2	25	1	29	29	0	1	0
				17	2	26	1	30.5	30.5	0	1	0
				21	2	26	2	30.5	0	30.5	0	1
				17	2	27	1	32	32	0	1	0
				20	2	29	1	33	33	0	1	0
				24	2	30	1	34.5	34.5	0	1	0
				17	2	30	1	34.5	34.5	0	1	0
				21	2	31	1	36.5	36.5	0	1	0
				24	2	31	2	36.5	0	36.5	0	1
				37					424.5	278.5	20	17
											NX→	20
											TX→	424.5
											U→	125.5
								Zcalc numerator	-44.5		Z calc →	-1.356186376
								Zcalc denominator	32.81259921			
											Z crit →	1.959963985

Figure 8.15: MANN-WHITNEY "U" test—large sample—download ch8-ex.xls

Wilcoxon Sign Rank Test - Two Populations
Matched Pairs
Ordinal Data

before	after	difference	abs(diff)	rank	pos diff	neg diff	pos rank*sign	neg rank*sign	
25	27	2	2	9.5	1	0	9.5	0	
36	32	-4	4	19.5	0	1	0	19.5	
24	25	1	1	3.5	1	0	3.5	0	
31	33	2	2	9.5	1	0	9.5	0	
20	22	2	2	9.5	1	0	9.5	0	
22	25	3	3	16	1	0	16	0	
28	27	-1	1	3.5	0	1	0	3.5	
23	25	2	2	9.5	1	0	9.5	0	
27	32	5	5	21	1	0	21	0	
21	28	7	7	22	1	0	22	0	
20	22	2	2	9.5	1	0	9.5	0	
21	21	0	0	1	0	0	0	0	
24	27	3	3	16	1	0	16	0	
30	29	-1	1	3.5	0	1	0	3.5	
24	22	-2	2	9.5	0	1	0	9.5	
25	27	2	2	9.5	1	0	9.5	0	
19	22	3	3	16	1	0	16	0	
25	24	-1	1	3.5	0	1	0	3.5	
21	24	3	3	16	1	0	16	0	
20	24	4	4	19.5	1	0	19.5	0	
20	23	3	3	16	1	0	16	0	
22	20	-2	2	9.5	0	1	0	9.5	
22 ← n							203	49	
								49 ← W	
					Z numerator	-72		-2.337526817 ← Z calc	← for large samples – off the table
					Z denominator	30.80178566			
								1.959963985 ← Z crit	

Figure 8.16: Wilcoxon—differences between two samples—download ch8-ex.xls

WILCOXON SIGNED RANK TEST for ONE POPULATION analysis can be used. The number of matched pairs differences are the "n" in the table.

The WILCOXON SIGN RANK TEST table of "critical" values is in the Appendix.

SUMMARY

You were introduced to two population hypothesis testing and its similarity to one population hypothesis testing. ALL the "rules" about hypothesis testing introduced in Chap. 7 are applicable. You were also introduced to the logic of matched pairs hypothesis testing.

You have been introduced to:

- The logic of independent and dependent populations (and samples)
- The copyable formula to analyze nominal data: the Chi-Square 2×2 and 2×n Contingency Tables for independent samples, and the Sign Test for dependent samples
- How to use OpenOffice to analyze ordinal data: the Mann-Whitney Test for independent samples, and the Wilcoxon Sign Rank Test for dependent samples

- How to interpret the various test results

EXERCISES

1. Assume that you are the public relations director for the county health department, and you want to publicize that getting a flu shot is worth the time, money, and effort regarding preventing the flu. You randomly selected records for 650 people and found these results:

	Got Flu Shot	Didn't Get Flu Shot
Had Flu	15	360
Didn't have flu	255	20

   a. State the null and alternative hypotheses
   b. Should you publicize the findings? Why or why not?

2. Let's use the same scenario as in No. 1, but this time we are concerned only with peoples' reactions. These results were found:

	Got Flu Shot	Didn't Get Flu Shot
Had Flu, saw doctor	10	240
Had Flu, didn't see doctor	35	400
Had Flu, missed work	5	600
Didn't have flu	755	20

   a. State the null and alternative hypotheses
   b. Should you publicize the findings? Why or why not?
   c. How many degrees of freedom are there in this analysis?

3. As the manager of a small company, you wonder if outside sales experience improves counter sales with respect to customer impressions. You randomly select ten employees who have had both kinds of sales, then ask customers to rate their impressions. (Customer impressions: -1=negative, 0=neutral, 1=positive)
   The data are in the spreadsheet ch8-data.xls

   a. State the null and alternative hypotheses
   b. Is there a relationship with respect to customer impressions? Explain
   c. Can this be a matched pairs environment? Explain.

4. Let's use the same scenario as with No. 3, except this time you are the manager of a large company, and this time randomly select 30 employees.
   The data are in the spreadsheet ch8-data.xls

a. State the null and alternative hypotheses

b. Is there a relationship with respect to customer impressions? Explain.

c. Can this be an independent samples environment? Explain.

5. Let's use the same scenario as with No. 3, except this time customers are asked to score salesperson effectiveness (higher scores mean more effective).
The data are in the spreadsheet ch8-data.xls

    a. State the null and alternative hypotheses

    b. Is there a relationship with respect to customer scores? Explain.

    c. Is this best as an independent samples or matched pairs environment? Explain.

6. Let's use the same scenario as with No. 5, except this time customers are asked to score 25 salespersons' effectiveness (higher scores mean more effective).
The data are in the spreadsheet ch8-data.xls

    a. State the null and alternative hypotheses

    b. Is there a relationship with respect to customer scores? Explain.

7. Assume that you are the operations manager of a large manufacturing company. A newly hired vice president wants to know if men and women are similarly represented in all types of manufacturing jobs: receiving, manufacturing, inventory, and shipping. What do you tell her?
The data are in the spreadsheet ch8-data.xls

8. Assume that you, as safety officer for a large company, have emphasized personal safety while on the job for the past year. You want to see if your efforts have had any effect. You randomly select 10 departments with virtually the same employee duties, and examine accident records for similar times for the departments.
The data are in the spreadsheet ch8-data.xls

    a. State the null and alternative hypotheses

    b. Are your efforts having any effect? Explain.

    c. Is this best as an independent samples or matched pairs environment? Explain.

    d. For what factor(s) might your analysis not be able to account?

Exercise data and solutions spreadsheets can be downloaded.

9 Three (or More) Populations

One way to think about play, is as the process of finding new combinations for known things...
Fred Rogers, children's TV personality

Note that with three or more populations there is no such thing as "matched pairs."

If you have three (or more) populations, you CANNOT use the procedures outlined in Chaps. 7 and 8. There are two reasons:

1. The procedures discussed in Chaps. 7 and 8 work ONLY for one and two populations.
2. Repeated analysis of the same samples results in the inflation of α, the probability of committing a Type I error (see the Introduction).

Therefore, if you have more than two populations you **MUST** use these procedure because they directly addresses both problems described above.

The logic of Three or More Population Hypothesis Testing is the same as with one or two population hypothesis testing: decision support.

HYPOTHESES

H_0: all population values of interest are the same ($\mu 1 = \mu 2 = \mu 3 = ...$) any number of populations can be examined
H_1: NOT H_0, that is **AT LEAST ONE** of the populations differs from at least one other population:
$\mu 1 \neq \mu 2 = \mu 3$
$\mu 1 = \mu 2 \neq \mu 3$
and so forth.... The applicable H_1 is never identified. But if H_0 is rejected, data inspection is warranted, so the applicable H_1 may be ascertained.

W. Beatty, *Decision Support Using Nonparametric Statistics*, SpringerBriefs in Statistics,
https://doi.org/10.1007/978-3-319-68264-8_9

α is (as before) an expression of how confident you want to be about making the correct decision regarding H_0 when conducting the hypothesis test

A primer on hypothesis testing can be found in the introduction.

NOMINAL DATA

CHI-SQUARED "n×m" CONTINGENCY TABLE

All that stuff you learned in Chap. 8 about "2×n" Chi-Squared Contingency Tables still applies.

The Chi-Squared "n×m" Contingency Table addresses three (or more) populations of interest: usually (but not necessarily) the "n" in the "n×m" designation. The values of "n" and "m" can be anything desired (as long as they are at least two), although "n" and/or "m" values over four are very difficult to inspect and interpret if H_0 is rejected.

Let's use the same new soft drink scenario introduced in Chap. 8, but there are now three categories of people, defined by age: young, middle aged, and older. There are five possible responses to the reaction to a taste test: Great Dislike, Dislike, Neutral, Like, Great Like. Now the question to be answered via analysis is whether taste test responses are the same (your H_0) for all three age categories. There are 3×5 response categories, so DF = 2 * 4 = 8. The spreadsheet is illustrated in Fig. 9.1.

Chi-Squared Analysis - Three Populations
Nominal Data

Observed Data	Young	Middle Age	Older	Totals
Great Dis	10	30	25	65
Dislike	25	30	30	85
Neutral	40	40	30	110
Like	30	30	40	100
Great Like	40	20	20	80
Totals	145	150	145	440

Expected Data	Young	Middle Age	Older	Totals
Great Dis	21.42045455	22.15909091	21.42045455	65
Dislike	28.01136364	28.97727273	28.01136364	85
Neutral	36.25	37.5	36.25	110
Like	32.95454545	34.09090909	32.95454545	100
Great Like	26.36363636	27.27272727	26.36363636	80
Totals	145	150	145	440

Chi-Square Calculation				Totals
Great Dis	6.088889559	2.774475524	0.598173378	9.461538462
Dislike	0.323736862	0.036096257	0.141181081	0.501014199
Neutral	0.387931034	0.166666667	1.077586207	1.632183908
Like	0.264890282	0.490909091	1.506269592	2.262068966
Great Like	7.053291536	1.939393939	1.536050157	10.52873563

This is the calculated Chi-Squared value. — 24.38554117

This is the probability of the calculated Chi-Squared value IF there is no difference between age groups. DF=8 Value returned by CHIDIST function. — 0.001974229

Figure 9.1: CHI-SQUARED test of independence with 3×5 table—download ch9-ex.xls

Figure 9.1 is difficult to read, so download ch9-ex.xls for complete access to the spreadsheet.

The p-value for the calculated Chi-Squared value of 24.385 is 0.001974, or 0.1974%. Since the p-value is less than (typically) $\alpha = 0.05$, H_0 CAN be rejected, and original data inspection and interpretation is warranted. The middle-age people appear to favor it slightly over the other two age groups.

Aside: does the above analysis influence your advertising decision? You rejected H_0, so the answer is "Yes" from that perspective. But inspection indicates such a small difference between the middle-aged group and the other two groups that further analysis, particularly a larger sample, is warranted. The lesson here: don't abandon your business knowledge—remember, statistics is nothing more than another source of knowledge.

Also, from the table in the Appendix, the "critical" value (df=8, α=0.05) is 15.51, so H_0 CAN be rejected, and that consistency should comfort you.

Here is another example, same scenario, except this time there are four regions of the country. The spreadsheet is illustrated in Fig. 9.2.

Chi-Squared Analysis - Four Populations
Nominal Data

	North	South	East	West	
	Observed Data				Totals
Great Dis	15	20	15	25	75
Dislike	20	20	20	30	90
Neutral	15	10	15	15	55
Like	40	50	40	40	170
Great Like	45	30	35	25	135
Totals	135	130	125	135	525

North	South	East	West	
Expected Data				Totals
19.28571429	18.57142857	17.85714286	19.28571429	75
23.14285714	22.28571429	21.42857143	23.14285714	90
14.14285714	13.61904762	13.0952381	14.14285714	55
43.71428571	42.0952381	40.47619048	43.71428571	170
34.71428571	33.42857143	32.14285714	34.71428571	135
135	130	125	135	525

Chi-Square Calculation				Totals
0.952380952	0.10989011	0.457142857	1.693121693	3.212535613
0.42680776	0.234432234	0.095238095	2.031746032	2.788224122
0.051948052	0.961704962	0.277056277	0.051948052	1.342657343
0.315592904	1.484378367	0.005602241	0.315592904	2.121166415
3.047619048	0.351648352	0.253968254	2.718400941	6.371636594
				15.83622009
				0.198853176

This is the calculated Chi-Squared value. → 15.83622009

This is the probability of the calculated Chi-Squared value IF there is no difference between age groups. DF=12 Value returned by CHIDIST function. → 0.198853176

Figure 9.2: CHI-SQUARED test of independence with 4×5 table—download ch9-ex.xls

The p-value for the calculated Chi-Squared value of 15.836 is 0.19885, or 19.885%. Since the p-value is greater than (typically) $\alpha = 0.05$, H_0 CANNOT be rejected, and original data inspection and interpretation is not warranted. You must continue to assume that all four regions have similar taste preferences regarding the new soft drink.

Also, from the χ^2 table in the Appendix, the "critical" value (df=12, α=0.05) is 21.03, so H_0 CANNOT be rejected. The consistency should comfort you.

ORDINAL DATA

KRUSKAL-WALLIS TEST

The Kruskal-Wallis (K-W) Test is an extension of the Mann-Whitney Test (see Chap. 8) that examines ranks from three (or more) populations to evaluate whether they all came from the same (or similar) population(s).

K-W determines ONLY whether at least one of the samples differs from the other samples.

The specific sample or samples that are different is/are not identified by K-W, nor is the amount of difference(s). If, however, H_0 is rejected, inspection is warranted, and that MAY suggest which population to favor or discard.

Let's continue the salesperson scenario introduced in Chap. 8, but this time you select five sales employees from the east, west, north, and south sales districts, and rank them as to sales effectiveness. The salespeople from each district are trained by that district's head salesperson, and their scores (however ascertained) are recorded. The question being asked is: are all four districts' rankings (and the training) the same? This is H_0. The spreadsheet is illustrated in Fig. 9.3.

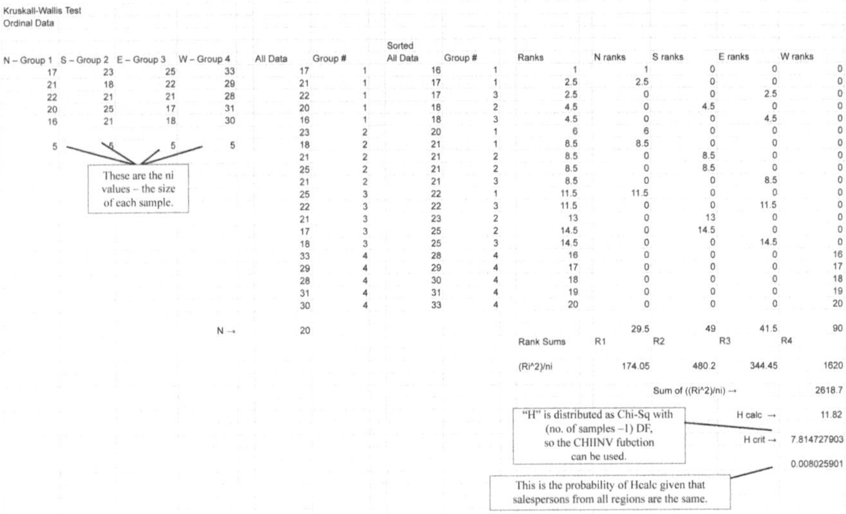

Kruskall-Wallis Test
Ordinal Data

N – Group 1	S – Group 2	E – Group 3	W – Group 4
17	23	25	33
21	18	22	29
22	21	21	28
20	25	17	31
16	21	18	30
5	5	5	5

These are the ni values – the size of each sample.

All Data	Group #	Sorted All Data	Group #	Ranks	N ranks	S ranks	E ranks	W ranks
17	1	16	1	1	1	0	0	0
21	1	17	1	2.5	2.5	0	0	0
22	1	17	3	2.5	0	0	2.5	0
20	1	18	2	4.5	0	4.5	0	0
16	1	18	3	4.5	0	0	4.5	0
23	2	20	1	6	6	0	0	0
18	2	21	1	8.5	8.5	0	0	0
21	2	21	2	8.5	0	8.5	0	0
25	2	21	2	8.5	0	8.5	0	0
21	2	21	3	8.5	0	0	8.5	0
25	3	22	1	11.5	11.5	0	0	0
22	3	22	3	11.5	0	0	11.5	0
21	3	23	2	13	0	13	0	0
17	3	25	2	14.5	0	14.5	0	0
18	3	25	3	14.5	0	0	14.5	0
33	4	28	4	16	0	0	0	16
29	4	29	4	17	0	0	0	17
28	4	30	4	18	0	0	0	18
31	4	31	4	19	0	0	0	19
30	4	33	4	20	0	0	0	20
N → 20					29.5	49	41.5	90
				Rank Sums	R1	R2	R3	R4
				(Ri^2)/ni	174.05	480.2	344.45	1620
						Sum of ((Ri^2)/ni) →		2618.7
							H calc →	11.82
							H crit →	7.814727903
								0.008025901

"H" is distributed as Chi-Sq with (no. of samples −1) DF, so the CHIINV fubction can be used.

This is the probability of Hcalc given that salespersons from all regions are the same.

Figure 9.3: KRUSKAL-WALLIS four population test—download ch9-ex.xls

The K-W Test H value is distributed similarly to the Chi-Squared distribution (d.f. = No. of samples − 1) so CHIINV and CHIDIST can be used.

The calculated H value of 11.82 has a p-value of 0.8% (from CHIDIST) if H_0 cannot be rejected, so the p-value for H_{calc} is less than (typical) $alpha = 0.05$. H_0 CAN be rejected.

Similarly, H_{crit} (from CHIINV) is 7.814, smaller than H_{calc} of 11.82, so H_0 CAN be rejected.

H_0 CAN be rejected, so all scores are not the same for all districts. Observed data inspection and analysis is warranted since H_0 was rejected. Cursory inspection indicates that the west region is most effective, so studying its methods are suggested.

Again, this spreadsheet looks complicated, but it's not. The "copy" and "sort" operations are identical to that used for the Mann-Whitney U Test in Chap. 8. This spreadsheet can be copied, so all the hard work has been done!

Here is an other example, same sales scenario, but this time from the east and west coasts, as well as the mid-west region. The spreadsheet is illustrated in Fig. 9.4.

Kruskall-Wallis Test
Ordinal Data

East Coast	Group #	West Coast	Group #	Mid-West	Group #		All Rankings	Group #	Sorted All Rankings	Group #	Ranking	East Coast Ranks	West Coast Ranks	Mid-West Ranks
15	1	16	2	14	3		15	1	11	3	1.5	0	0	1.5
19	1	18	2	11	3		19	1	11	3	1.5	0	0	1.5
20	1	22	2	17	3		20	1	14	3	3	0	0	3
18	1	21	2	15	3		18	1	15	1	4.5	4.5	0	0
21	1	24	2	11	3		21	1	15	3	4.5	0	0	4.5
22	1	23	2	34	3		22	1	16	2	6	0	6	0
31	1	28	2	36	3		31	1	17	3	7	0	0	7
							16	2	18	1	8.5	8.5	0	0
7		7		7			18	2	18	2	8.5	0	8.5	0
							22	2	19	1	10	10	0	0
							21	2	20	1	11	11	0	0
							24	2	21	1	12.5	12.5	0	0
							23	2	21	2	12.5	0	12.5	0
							28	2	22	1	14.5	14.5	0	0
							14	3	22	2	14.5	0	14.5	0
							11	3	23	2	16	0	16	0
							17	3	24	2	17	0	17	0
							15	3	28	2	18	0	18	0
							11	3	31	1	19	19	0	0
							34	3	34	3	20	0	0	20
							36	3	36	3	21	0	0	21
						N	21				Sums	80	92.5	58.5
									Sums^2/ni →			914.2857143	1222.321429	488.8928571
												Sum of (Sums^2/ni) →		2625.5
													H calc →	2.194805195
													H crit →	5.991464547
													H p-val →	0.333736808

Figure 9.4: K-W three region test—download ch9-ex.xls

In this example, the calculated H value of 2.194 has a p-value of 0.333 if H_0 cannot be rejected. This p-value is greater that (typical) $\alpha = 0.05$, so H_0 CANNOT be rejected. Similarly, H_{crit} is 5.991, greater than H_{calc} of 2.194, so H_0 CANNOT be rejected. You must continue to assume that ranks are the same for all three districts.

SUMMARY

You have been introduced to three (or more) population hypothesis testing.

You have been introduced to:

- The logic of analyzing three (or more) populations
- The formulation of the null (H_0) and alternative (H_1) hypotheses
- How to use OpenOffice to analyze nominal data: the Chi-Square "n×m" Contingency Tables

- How to use OpenOffice to analyze ordinal data: the Kruskal-Wallis Test
- How to interpret the various test results

EXERCISES

1. Assume that you are the marketing manager for a national chicken restaurant that has five styles of chicken: Regular, Mild, Spicy, Hot, and Hurt. You want to know if there is a relationship between chicken styles and customer age groups: younger, middle age, and older. The data are:

	Regular	Mild	Spicy	Hot	Hurt
Younger	10	25	30	50	15
Middle Age	20	30	35	30	10
Older	50	45	40	30	10

   (a) State the null hypothesis (H_0) and the alternative hypothesis (H_1)
   (b) Is this a "3×5" or a "5×3" situation? Explain.
   (c) Can H_0 be rejected? Explain.
   (d) Is observed data inspection and interpretation warranted? Explain.
   (e) What is the most popular style of chicken? The least popular?

2. Assume that you are the marketing manager for a national fast-food restaurant that offers a "kiddie meal" that can contain a toy or a snack in addition to the child sized meal, or the meal alone. You want to know if kiddies (pre-school, kindergarten, school age [1st, 2nd, 3rd grade]) select the meal because of the toy, the snack, or the meal. The data are:

	Toy	Snack	Meal
Preschool	50	70	21
Kindergarten	30	50	20
School Age	25	32	12

   (a) State the null hypothesis (H_0) and the alternative hypothesis (H_1)
   (b) Can (H_0) be rejected? Explain.
   (c) Is observed data inspection and interpretation warranted? Explain.

3. Assume that you are the manager of a large, nationwide company, and want to investigate customer satisfaction with respect to regions of the country. You randomly select 15 customers from five country regions: east coast, west coast, deep south, midwest, and mountain are, and solicit satisfaction scores: from 5=very satisfied, to 1=very dissatisfied. The data are in ch9-data.xls

   (a) State the null hypothesis (H_0) and the alternative hypothesis (H_1)

   (b) Can H_0 be rejected? Explain.

   (c) Is observed data inspection and interpretation warranted? Explain.

   (d) What is the most satisfied region? The least satisfied?

4. Now assume that you are the manager of a fast-food restaurant that is doing a brisk business. You are interested in investigating if customer satisfaction (1 to 10, higher score means more satisfied) is related to age groups. You create three age groups (<25 years of age, 25–55 years of age, >55 years of age), then ask 35 randomly selected customers to compete a survey.
   The data are in ch9-data.xls

   (a) State the null hypothesis (H_0) and the alternative hypothesis (H_1)

   (b) Can (H_0) be rejected? Explain.

   (c) Is observed data inspection and interpretation warranted? Explain.

   (d) What is the most satisfied age group? The least satisfied?

Exercise data and solutions spreadsheets can be downloaded.

10
Association

Association does not mean causation
Dr. Malcolm Kendrick

Measures of association provide a means of expressing the degree or amount of association between two variables, typically named X and Y. Many measures of association are constructed to return values between ± 1 (perfect association) and 0 (no association). The closer to 1, the "stronger" the association. Some return values between -1 and 1, so the nature of the association is expressed. Negative values indicate a negative or inverse relationship (Y values decrease as X values increase), while positive values indicate a positive relationship (Y values increase as X values increase). Hypothesis tests are possible for many of the measures of association. They begin by hypothesizing that there is NO relationship between the two variables, or that the measure of association equals 0.

REMEMBER:
Causation is a **managerial interpretation** of association.
There must be some logical reason for the association. All this analysis does is quantify, or hang a number on, the association.
Also remember what Dr. Kendrick says: "Association does not mean causation."

NOMINAL DATA

CHI-SQUARED 2×2 CONTINGENCY TABLE

ϕ is a product-moment coefficient of correlation (association), and is a variation of Pearson's ρ when both variables are dichotomous. ϕ is designed to measure the association between two dichotomous variable distributions. The Pearson product-moment correlation coefficient can be simplified if both the X and Y variables are nominal and dichotomous, and a Chi-Squared 2×2 Contingency Table can be used.

W. Beatty, *Decision Support Using Nonparametric Statistics*,
SpringerBriefs in Statistics,
https://doi.org/10.1007/978-3-319-68264-8_10

The Measures of Association Phi (ϕ) can be calculated using this formula: $\phi = (\mathrm{BC} - \mathrm{AD})/\sqrt{((A+B)(C+D)(A+C)(B+D))}$,
where A, B, C, and D are the values in the following table:

A	B
C	D

Let's use the Men/Women, Like/Don't Like soft drink scenario from Chap. 8. Here is the data:

	Men	Women
Like	10 (A)	50 (B)
Don't Like	30 (C)	40 (D)

The spreadsheet is illustrated in Fig. 10.1.

Chi-Squared 2x2 Contingency Table
Nominal Data
Measure of Association

	Men	Women		
Like	10	50	A	B
Don't Like	30	40	C	D
	Phi →	0.282889475		

Φ = (BC - AD)/sqrt((A+B)(C+D)(A+C)(B+D)),

Figure 10.1: 2×2 Contingency table measure of association—download ch10-ex.xls

The ϕ value can be evaluated thusly (note: this is a "rule of thumb" evaluation)

ϕ Value	Association Strength
−0.8 to −1.0 or 0.8 to 1.0	Very Strong
−0.5 to −0.8 or 0.5 to 0.8	Strong
−0.3 to −0.5 or 0.3 to 0.5	Moderate
−0.1 to −0.3 or 0.1 to 0.3	Weak
−0 to −0.1 or 0 to 0.1	Very Weak or Non-existent

Based on this sample, is any association between sex and taste preference exhibited? The ϕ value of 0.282889 suggests a weak association between sex and taste preference.

ϕ may be used if observations are grouped into two arbitrarily (but clearly) defined groups. For example, say you want to see if there is an association between sales effectiveness and college. First, if "sales effectiveness" can logically be divided into two groups (say High and Low), then it can be used. Similarly, if "college" can logically be divided into two groups (say bachelor's degree as the division point—have or don't have), then it can also be used to calculate ϕ. The key concept is that "arbitrary" be clearly, consistently, logically defined before any data are collected.

CHI-SQUARED "n×m" CONTINGENCY TABLE

A ϕ value can be calculated for contingency tables larger than 2×2. Here $\phi = \sqrt{x}/\text{n}$ where "n" is the total sample size and "x" is the calculated value of χ^2.

Let's again use the new soft drink scenario. "Greatly Like, Like, Neutral, Don't Like, and Greatly Don't Like" form 5 rows, while "North, South, East, and West" form 4 columns. The spreadsheet is illustrated in Fig. 10.2.

Chi-Squared Analysis - Four Populations by Five Categories
Nominal Data
Measure of Association

Observed Data	North	South	East	West	Total
Great Dis	15	10	15	10	50
Dislike	20	15	20	15	70
Neutral	15	10	15	5	45
Like	40	50	45	50	185
Great Like	45	60	50	50	205
Total	135	145	145	130	555

Expected Data	North	South	East	West	Total
Great Dis	12.16216216	13.06306306	13.06306306	11.71171171	50
Dislike	17.02702703	18.28828829	18.28828829	16.3963964	70
Neutral	10.94594595	11.75675676	11.75675676	10.54054054	45
Like	45	48.33333333	48.33333333	43.33333333	185
Great Like	49.86486486	53.55855856	53.55855856	48.01801802	205
Total	135	145	145	130	555

This is the "n" value to use when calculating Phi

Chi-Sq Calculation					Total
Great Dis	0.662162162	0.718235477	0.287200994	0.25017325	1.917771883
Dislike	0.519090519	0.591243953	0.160209471	0.118923869	1.389467812
Neutral	1.501501502	0.262503883	0.894687791	2.912335412	5.571028588
Like	0.555555556	0.057471264	0.229885057	1.025641026	1.868552903
Great Like	0.474620962	0.774706582	0.23643913	0.081807887	1.567574562

This is the Chi-Sq value to use when calculating Phi

Chi-Sq → 12.31439575

Phi → 0.148956707

Figure 10.2: "n×m" Contingency table measure of association—download ch10-ex.xls

As can be seen, ϕ has a value of 0.149, so a weak association between taste preference and country region is exhibited. This is "good" because no association between taste preference and country region is ideal. Observed data, however, shows that all country regions dislike the new soft drink by very similar margins. There may be an association, but it's not what you want to see.

Note: ϕ, when used in a comparison environment, should only be compared to other ϕ values calculated from tables with similar dimensions.

For Chi-Square Contingency Tables that exceed 2×2, Cramer's V (also known as Cramer's ϕ) is the **preferred** statistic because it takes into account the dimensions of the contingency tables. Cramer's V has a maximum value of 1 when there is a very strong relationship between X and Y, and a value of 0 if X and Y are unrelated.

It is calculated as $\sqrt{x}/\text{nt}$ where x is the calculated value of χ^2, n is the total sample size, and t is the (lesser of the contingency table dimensions) − 1.

The same soft drink scenario as above is again used. The Cramer's V calculation is illustrated in the spreadsheet shown in Fig. 10.3.

Cramers V - Four Populations by Five Categories
Nominal Data
Measure of Association

Observed Data	North	South	East	West	Total
Great Dis	15	10	15	10	50
Dislike	20	15	20	15	70
Neutral	15	10	15	5	45
Like	40	50	45	50	185
Great Like	45	60	50	50	205
Total	135	145	145	130	555

Expected Data	North	South	East	West	Total
Great Dis	12.16216216	13.06306306	13.06306306	11.71171171	50
Dislike	17.02702703	18.28828829	18.28828829	16.3963964	70
Neutral	10.94594595	11.75675676	11.75675676	10.54054054	45
Like	45	48.33333333	48.33333333	43.33333333	185
Great Like	49.86486486	53.55855856	53.55855856	46.01801802	205
Total	135	145	145	130	555

This is the "n" value to use when calculating Cramer's V

Chi-Sq Calculation					Total
Great Dis	0.662162162	0.718235477	0.287200994	0.25017325	1.917771883
Dislike	0.519090519	0.591243953	0.160209471	0.118923869	1.389467812
Neutral	1.501501502	0.262503883	0.894687791	2.912335412	5.571028588
Like	0.555555556	0.057471264	0.229885057	1.025641026	1.868552903
Great Like	0.474620962	0.774706582	0.23643913	0.081807887	1.567574562

This is the Chi-Sq value to use when calculating Cramer's V

Chi-Sq → 12.31439575

V → 0.001580715

Contingency Table Dimensions	
Columns	4
Rows	5
Lesser	4

This is the "n" value to use when calculating Cramer's V

Figure 10.3: Cramer's V measure of association—download ch10-ex.xls

The Cramer's V value of 0.00158 indicates a nonexistent association between taste preference and country region. There is again a "however." Data inspection indicates that a large percentage of the tasters in the south "Like" or "Greatly Like" the new soft drink. Also, people in all regions "Greatly Like" the new soft drink.

ORDINAL DATA

SPEARMAN'S RANK CORRELATION COEFFICIENT

Spearman's rank correlation coefficient, Rho (ρ) measures the nature (positive or negative) and strength (very strong to non-existent) of linear relationship or association between two variables. It is calculated by using the Pearson correlation method applied to the ranks of the values of the two variables. This is done in order to alleviate problems of outliers, unequal variances, and non-normality of either or both variables that can greatly affect the Pearson correlation calculation.

Calculating ρ is quite easy to do. First, rank the data for both variables, usually named X and Y. The ranking procedure was introduced in Chap. 7. Once that is accomplished, use the OpenOffice PEARSON function. It's that easy!

Nature: Values of ρ less than zero signify a negative relationship, or the ranks of one variable decreases as the ranks of the other variable increases. Values of Rho greater than zero signify a positive relationship, or the ranks of one variable increases as the ranks of the other variable increases. The terms "negative" and "positive" are just names.

Strength: The value of ρ can be from -1 to 1. Values closer to -1 and 1 indicate stronger relationships, while values close to 0 indicate no relationship. There is no "official" rating of the value of ρ, but this table provides a "rule of thumb" for interpretation.

ρ Value	Association Strength
-0.8 to-1.0 or 0.8 to 1.0	Very Strong
-0.5 to -0.8 or 0.5 to 0.8	Strong
-0.3 to -0.5 or 0.3 to 0.5	Moderate
-0.1 to -0.3 or 0.1 to 0.3	Weak
-0 to -0.1 or 0 to 0.1	Very Weak or Non-existent

As an example, assume that you want to examine the association between salesperson effectiveness scores (X) and customer satisfaction scores (Y). Both sets of scores utilized the Likert scale. Is there a relationship? If so, what is its nature and strength? The spreadsheet is illustrated in Fig. 10.4.

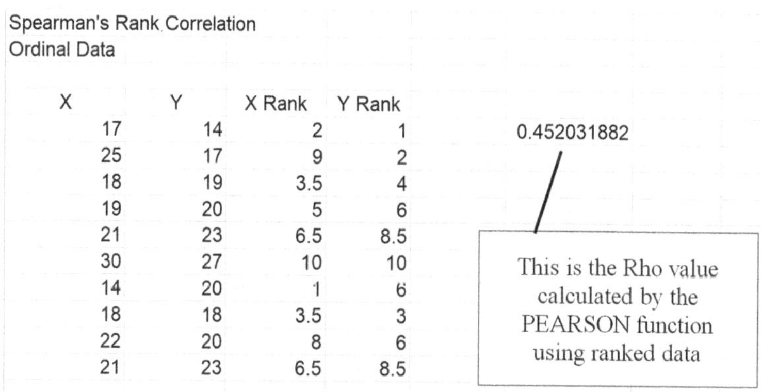
Spearman's Rank Correlation
Ordinal Data

X	Y	X Rank	Y Rank
17	14	2	1
25	17	9	2
18	19	3.5	4
19	20	5	6
21	23	6.5	8.5
30	27	10	10
14	20	1	6
18	18	3.5	3
22	20	8	6
21	23	6.5	8.5

Figure 10.4: Spearman's measure of association—download ch10-ex.xls

Hypothesis testing is possible with the Spearman's ρ rank correlation coefficient. H_0 is that $\rho = 0$, or that no relationship exists. A table of "critical values" is provided in the Appendix. Assuming $\alpha = 0.05$ (the typical value), compare the calculated ρ value of 0.452 with the "critical" ρ value (for n = 10) of 0.648 in the table in the Appendix. Since the calculated ρ value of 0.452 is less than the "critical" value of 0.648, H_0 CANNOT be rejected. An association between scores cannot be determined based on this sample.

For sample sizes (n) greater than 20 pairs of values, $\rho * \sqrt{x}$ [where x = $((n-2)/(1-\rho^2))$] is approximately distributed as t with DF = n − 2, so the OpenOffice function TINV can be used.

Assume the same scenario as above, but this time you randomly select 25 salespeople and customers. The spreadsheet is illustrated in Fig. 10.5.

Spearman's Rank Correlation
Ordinal Data
Sample Size > 20

X	Y	X Rank	Y Rank		
17	14	2	1	Rho →	0.588326558
25	17	16	3		
18	19	4	5		
19	20	6	7		
21	23	8	13.5		3.489281814
30	27	22.5	20		
14	20	1	7		
18	18	4	4		
22	20	11	7		2.06865761
30	21	22.5	10.5		
21	26	8	17.5		
25	28	16	22		
27	30	19	23		
27	21	19	10.5		
28	31	21	24		
22	27	11	20		
18	15	4	2		
31	24	24	15.5		
33	27	25	20		
24	26	13.5	17.5		
27	21	19	10.5		
24	32	13.5	25		
22	21	11	10.5		
25	24	16	15.5		
21	23	8	13.5		

25 ← Sample Size

This is the Rho value returned By the PEARSON function

This is the "t" value calculated using the formula Rho*sqrt((n-2)/(1-Rho^2))

This is the "critical t" value returned by the TINV function with n-2 DF

Figure 10.5: Spearman's measure of association—more than 20 pairs—download ch10-ex.xls

As can be seen, the calculated value of 3.489 is greater that t_{crit} value of 2.068, so H_0 CAN be rejected. An association between scores can be determined based on this sample. Make the managerial decision accordingly. Note that the analysis did not make the decision. Rather, it provided information to support the decision.

For sample sizes (n) greater than 40 pairs of values, $\rho * \sqrt{(n+1)}$ is approximately normally distributed, so NORMINV can be used.

Assume the same scenario as above, but this time you randomly select 45 salespeople and customers. The spreadsheet is illustrated in Fig. 10.6.

As can be seen, the calculated value of 1.663 is not greater that (the absolute value of) Z_{crit} value of 1.96, so H_0 CANNOT be rejected. Even though an association (a weak one) between scores can be determined based on this sample, H_0 CANNOT be rejected. Make the managerial decision accordingly. Note that the analysis did not make the decision. Rather, it provided information to support the decision.

WARNING!

There may be an association between variables even though the null hypothesis (H_0) stating that no relationship exists cannot be rejected. So, be very careful drawing conclusions about association, or about acting upon conclusions!

SUMMARY

You have been introduced to association and hypothesis testing.

- The concept of association, and that association doesn't necessarily indicate causation
- Hypothesis testing for association, and the formulation of the null (H_0) and alternative (H_1) hypotheses
- How to use OpenOffice to analyze nominal data: the Chi-Square 2×2 and 2×n Contingency Tables, the calculation of ρ and Cramer's V
- How to use OpenOffice to analyze ordinal data: the Spearman's Rank Correlation Coefficient (ρ)
- How to interpret the various tests and hypothesis test results

---

Spearman's Rank Correlation
Ordinal Data
Sample Size > 40

X	Y	X Rank	Y Rank		
17	14	2	1	0.245167553	This is the Rho value returned By the PEARSON function
25	17	27	3		
18	19	4	5		
19	20	6.5	7		
21	23	12	13.5	1.662807246	This is the "Z" value calculated using the formula Rho*sqrt(n+1)
30	27	41	22		
14	20	1	7		
18	18	4	4		
22	20	17.5	7		
30	21	41	10.5	-1.959963985	This is the "critical Z" value returned by the NORMINV function
21	26	12	18		
25	28	27	26		
27	30	34	32.5		
27	21	34	10.5		
28	31	37	36.5		
22	27	17.5	22		
18	15	4	2		
31	24	43	15.5		
33	27	44.5	22		
24	26	23	18		
27	21	34	10.5		
24	32	23	40.5		
22	21	17.5	10.5		
25	24	27	15.5		
21	30	12	32.5		
25	29	27	29		
30	34	41	44		
27	28	34	26		
29	27	38.5	22		
19	31	6.5	36.5		
21	32	12	40.5		
21	26	12	18		
23	29	20.5	29		
26	31	30.5	36.5		
22	33	17.5	43		
27	30	34	32.5		
33	27	44.5	22		
20	32	8	40.5		
26	30	30.5	32.5		
23	28	20.5	26		
25	32	27	40.5		
21	35	12	45		
24	31	23	36.5		
29	29	38.5	29		
21	23	12	13.5		

45 ← Sample Size

Figure 10.6: Spearman's measure of association—more than 40 pairs—download ch10-ex.xls

EXERCISES

1. Assume that you are a researcher for a large advertising company. One of the company's primary customers is a tobacco company. In order to maximize advertising effectiveness, the tobacco company asks you to see

if there is an association between smoking and the sex. You randomly observe 2505 people and record the results. Calculate ϕ. The data are:

	Men	Women
Smoke	530	190
Don't Smoke	1025	760

Is there an association between smoking and sex? If yes, what is it, and how would you interpret it for the tobacco company?

2. Using the same scenario as in No. 1, now expand the smoking scenario to four age groups: teens, 20–35, 36–50, over 50. This time only smokers are considered. The data are:

	Teens	20–35	36–50	Over 50
Men	220	300	250	110
Women	125	200	135	85

Is there an association between sex and age among smokers? If yes, what is it, and how would you interpret it for the tobacco company?

3. Assume the same scenario as in No. 2, but with different data:

	Teens	20–35	36–50	Over 50
Men	310	600	120	40
Women	115	300	70	35

Is there an association between sex and age among smokers? If yes, what is it, and how would you interpret it for the tobacco company?

4. Use the same data as in No. 3, but calculate Cramer's V. Is there any change in interpretation?

5. Now consider that you are studying age and daily smoking habit. The data are:

	Less than one pack	One pack	Two packs	Over two packs
Teens	10	50	75	90
20–35	25	45	70	80
36–50	75	35	25	15
Over 50	130	60	20	10

Is there an association between daily smoking habit and age? If yes, what is it, and how would you interpret it for the tobacco company?

6. Assume that you are the owner of a fast-food restaurant, and want to know if there is an association between employee effort (X) and customer satisfaction (Y). You ascertain median effort and customer satisfaction for 20 days. Is there as association? If so, what is it? Is the association significant (α=0.05)? The data are in ch10-data.xls
7. Let's use the same scenario as in No. 6, but for a different restaurant in a different part of the city. This time 36 days are observed. Is there as association? If so, what is it? Is the association significant (α=0.05)? The data are in ch10-data.xls
8. Let's use the same scenario as in No. 6, but for a different restaurant in a different city. This time 50 days are observed. Is there as association? If so, what is it? Is the association significant (α=0.05)? The data are in ch10-data.xls

Exercise data and solutions spreadsheets can be downloaded.

Appendix

SIGN Test

If the calculated "x" value is *less than or equal to* the table value, H_0 CAN be rejected

n	"critical" x value, α= .05	"critical" x value, α= .01
6	0	
7	0	
8	0	0
9	1	0
10	1	0
11	1	0
12	2	1
13	2	1
14	2	1
15	3	2
16	3	2
17	4	2
18	4	3
19	4	3
20	5	3
21	5	4
22	5	4
23	6	4
24	6	5
25	7	5
26	7	6
27	7	6
28	8	6
29	8	7
30	9	7

Values generated by author. Values verified from tables at http://www.social-science.co.uk/

W. Beatty, *Decision Support Using Nonparametric Statistics*,
SpringerBriefs in Statistics,
https://doi.org/10.1007/978-3-319-68264-8

Wilcoxon SIGN-RANK Test

If the calculated "W" value is *less than or equal to* the table value, H_0 CAN be rejected

n	"critical" W value, $\alpha = .05$	"critical" W value, $\alpha = .01$
6	0	
7	2	
8	3	0
9	5	0
10	8	3
11	10	5
12	14	7
13	17	9
14	21	12
15	25	16
16	29	19
17	35	23
18	40	27
19	46	32
20	52	37
21	58	44
22	65	47
23	73	54
24	81	61
25	89	68
26	98	75
27	107	83
28	116	91
29	126	101
30	136	106

Values generated by author. Values verified from tables at http://www.social-science.co.uk/

Chi-Squared (χ^2) Contingency Table Critical Values

If the calculated χ^2 value is GREATER than or equal to the table value, H_0 CAN be rejected

d.f.	α.= .05	α.= .01	α.= .001
1	3.84	6.64	10.83
2	5.99	9.21	13.82
3	7.82	11.36	16.27
4	9.49	13.28	18.47
5	11.07	15.09	20.52
6	12.59	16.81	22.49
7	14.07	18.48	24.32
8	15.51	20.10	26.13
9	16.92	21.67	27.88
10	18.31	23.21	29.59
11	19.68	24.73	31.26
12	21.03	26.22	32.91
13	22.36	27.69	34.53
14	23.69	29.14	36.12
15	25.00	30.58	37.70
16	26.30	32.00	39.25
17	27.59	33.41	40.79
18	28.87	34.81	42.31
19	30.14	36.19	43.82
20	31.41	37.57	45.32
25	37.65	44.31	52.62
30	43.77	50.89	59.70
40	55.76	53.69	73.40
50	57.51	76.15	86.66
100	124.34	135.81	149.45

Values from OpenOffice CHIINV function. Values verified from tables at http://www.itl.nist.gov/div898/handbook/eda/section3/eda3674.htm

Mann-Whitney "U" Test

α=0.05

If the calculated "U" value is less than or equal to the table value, H_0 CAN be rejected

	n_1 →	2	3	4	5	6	7	8	9	10	11	12	13	14	15	16	17	18	19	20
n_2 ↓																				
5			0	1	2															
6			1	2	3	5														
7			1	3	5	6	8													
8		0	2	3	6	8	10	13												
9		0	2	4	7	9	12	15	17											
10		0	3	5	8	11	14	17	20	23										
11		0	3	6	9	13	16	19	23	26	30									
12		1	3	7	11	14	18	22	26	29	33	37								
13		1	4	8	12	16	20	24	28	33	37	41	45							
14		1	5	9	13	17	22	26	31	36	40	45	50	54						
15		1	5	10	14	19	24	29	34	38	44	49	54	59	64					
16		1	6	11	16	21	26	31	36	42	48	53	59	64	69	74				
17		2	6	12	17	22	28	34	39	45	51	57	63	69	75	80	86			
18		2	6	13	18	24	30	36	42	49	55	61	67	73	80	86	93	99		
19		2	7	13	19	26	31	39	45	52	58	65	72	78	85	92	99	106	112	
20		2	8	14	20	27	34	41	48	55	62	68	76	83	90	98	104	112	119	127

α=0.01

If the calculated "U" value is less than or equal to the table value, H_0 CAN be rejected

	n_1 →	2	3	4	5	6	7	8	9	10	11	12	13	14	15	16	17	18	19	20
n_2 ↓																				
6				0	1	1														
7				0	1	3	4													
8				1	2	4	6	8												
9			0	1	3	5	7	9	11											
10			0	2	4	6	9	11	13	16										
11			0	2	5	7	10	13	16	18	21									
12			1	3	6	9	12	15	18	21	24	28								
13			1	3	7	10	13	17	20	23	27	31	34							
14			1	4	8	11	15	19	23	26	30	34	39	41						
15			1	5	8	12	16	20	24	28	33	37	41	47	52					
16			2	5	10	13	17	23	26	32	36	42	45	51	55	59				
17			2	6	10	14	19	24	29	35	39	44	54	54	59	64	70			
18			2	6	11	16	20	25	31	37	43	47	52	63	70	86	75	81		
19		0	3	7	12	17	22	28	34	39	45	51	57	63	69	75	81	87	93	
20		0	3	8	13	18	24	31	36	42	48	54	61	67	74	78	85	92	95	106

Values generated by author. Values verified from tables at http://www.social-science.co.uk/

Spearman Rank Correlation Coefficient Critical Values

If the calculated ρ value is **GREATER than or equal to** the table value, H_0 CAN be rejected

No. Pairs	α=.05	α=.01
5	1.0000	
6	0.8857	1.0000
7	0.7857	0.9286
8	0.7381	0.8810
9	0.6833	0.8167
10	0.6485	0.7939
11	0.6182	0.7546
12	0.5874	0.7273
13	0.5604	0.7033
14	0.5429	0.6791
15	0.5179	0.6536
16	0.5000	0.6206
17	0.4853	0.6177
18	0.4758	0.6037
19	0.4579	0.5842
20	0.4466	0.5684

Values generated by author. Values verified from tables at http://www.social-science.co.uk/

References

Chapter 7—One Population Hypothesis Testing

http://www.itl.nist.gov/div898/handbook/eda/section3/eda35d.htm
http://www.csun.edu/~mr31841/documents/thesigntest.pdf—page 5
http://mlsc.lboro.ac.uk/resources/statistics/wsrt.pdf

Chapter 8—Two Population Hypothesis Testing

http://changingminds.org/explanations/research/analysis/chi-square.htm
http://www.csun.edu/~mr31841/documents/thesigntest.pdf
http://www.sussex.ac.uk/Users/grahamh/RM1web/MannWhitneyHandout%202011.pdf
http://mlsc.lboro.ac.uk/resources/statistics/wsrt.pdf

Chapter 9—Three or More Population Hypothesis Testing

http://changingminds.org/explanations/research/analysis/chi-square.htm
http://vassarstats.net/textbook/ch14a.html
people.math.carleton.ca/~smills/2009-10/STAT4504/k-wtest.ppt

Chapter 10—Association

http://uregina.ca/~gingrich/ch11a.pdf
http://www.andrews.edu/~calkins/math/edrm611/edrm13.htm
http://www.people.vcu.edu/~pdattalo/702SuppRead/MeasAssoc/NominalAssoc.html
http://www.acastat.com/Statbook/chisqassoc.htm
http://changingminds.org/explanations/research/analysis/cramers_v.htm
http://www.mei.org.uk/files/pdf/Spearmanrcc.pdf
http://www.harding.edu/sbreezeel/460%20Files/Statbook/CHAPTER14.pdf

OpenOffice HELP

OPENOFFICE HELP

You can find many references on the Internet about how the OpenOffice (and Excel) statistics are inferior to SPSS, SAS, Minitab, and others. For serious statistics, they are correct. However, as an introductory teaching tool, OpenOffice is fine.

OpenOffice statistics functions and Microsoft Excel statistics functions are very similar. They are exactly alike in most cases. So, any Microsoft Excel links that are helpful can also apply to OpenOffice.

Here are some Excel specific links that you may find helpful with OpenOffice.

Link 1—Excel For Statistical Data Analysis:
http://home.ubalt.edu/ntsbarsh/excel/Excel.htm

Link 2—Excel Tutorial for Statistical Data Analysis;
http://www.stattutorials.com/EXCEL/index.html
note: the Excel Toolpak is NOT required to examine these explanations

OpenOffice itself has a Help section as well.

Spreadsheet Downloads

Go to: http://www.spreadsheet-download.co.nf to download ALL the spreadsheets.

Spreadsheet Tutorial

Quick and Dirty Spreadsheet Tutorial

DOWNLOAD and INSTALLATION

Windows Version

The web address where OpenOffice can be downloaded (free) is:
http://www.openoffice.org/download/
Click on the "down arrow" to start the download. Click on the "Run" button to start the installation.

NOTE: The OpenOffice spreadsheet (Calc) is part of the Apache OpenOffice suite. Only the spreadsheet can be installed if desired.

The entire installation process should take about 2 min and consume about 440 MB of disc space. The installation process is pure "Windows." A desktop icon will be created for you. You can, if you want, save the OpenOffice installation file, then run it a later time to accomplish the installation.

Apple Mac Version

Go here to download an Apple version: https://www.openoffice.org/porting/mac/

HELP

The OpenOffice spreadsheet comes with a very extensive Help facility. For this reason the Q and D Tutorial will focus on how to enter text, numbers, mathematical formulae, and statistical formulae into specific cells (Fig. A.1).

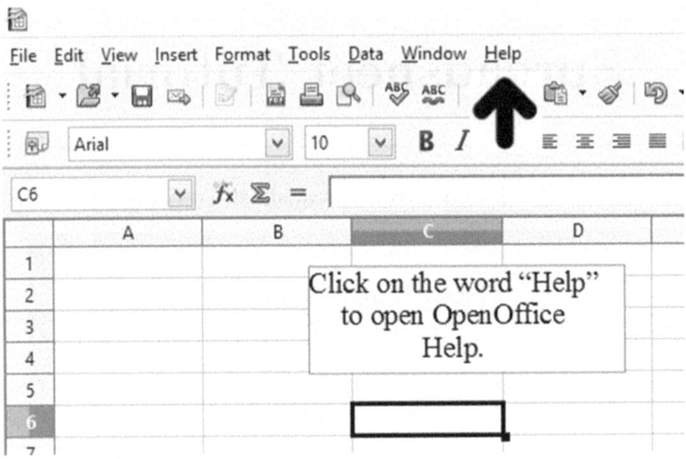

Figure A.1: Accessing OpenOffice HELP

COLUMNS, ROWS and CELLS

EVERY location within a spreadsheet, called a cell, **MUST** have a unique address name. This is accomplished through the COLUMN, ROW address naming convention. Look at the illustration in Fig. A.2.

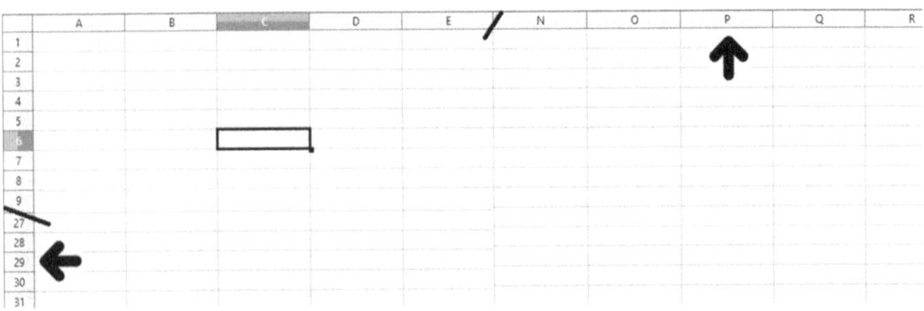

Figure A.2: Column, row naming of cells

Columns "A" through "P" are illustrated here. (by default, OpenOffice provides through column IV [256 columns]). Columns are referred to by letters.

Rows 1 through 29 are illustrated here. (by default, OpenOffice provides through row 65536). Rows are referred to by numbers.

In the illustration shown in Fig. A.2, cell (specific location) C6 is the "selected" or "active" cell. You know this because (1) cell C6 has a heavy outline around it, (2) column "C" and row "6" are highlighted in blue across the top and left side, and (3) the address of the selected cell appears in the upper left corner of the spreadsheet.

ANYTHING you type will go into the selected cell. At the same time, what you type will appear on both the selected cell and the "Input" line. You must, when your input is complete, press the "ENTER" key finalize input (Fig. A.3).

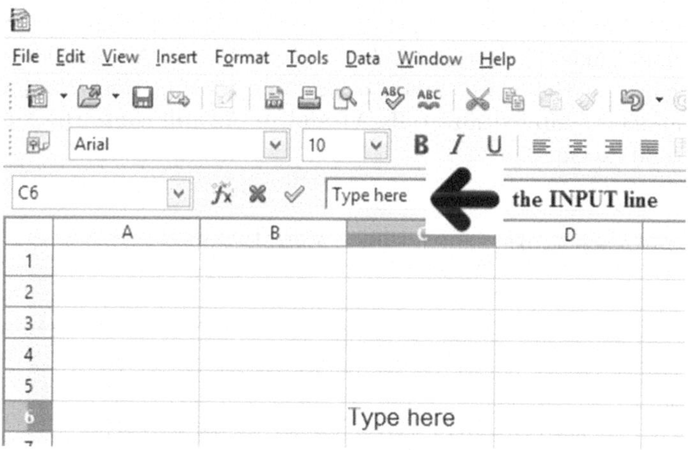

Figure A.3: The INPUT line

You can use your keyboard movement arrow keys (but not the 'ENTER' key) to move around the spreadsheet and to select other cells. **But you can select only one cell at a time.**

ENTERING MATHEMATICAL FORMULAE

One of the powers of any spreadsheet is its ability to accept mathematical formulae and to automatically recalculate formulae values as conditions change. For example, look at the illustration in Fig. A.4.

Cells A1 and A2 have (any) numbers entered. Cell A3 has the mathematical formula =A1 + A2 in it. That means that the sum of the numbers in cells A1 and A2 will be automatically displayed in cell A3. This is because cell

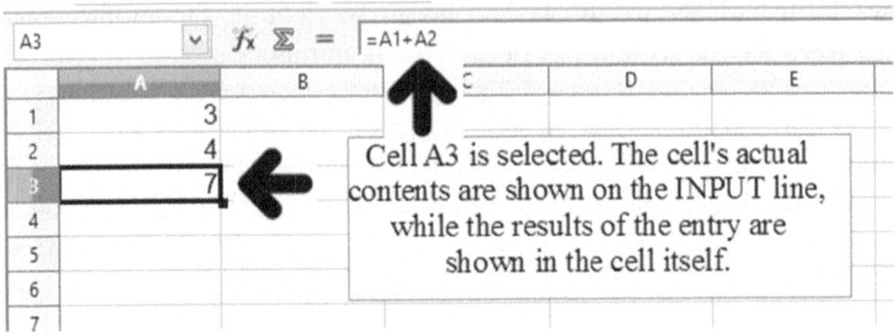

Figure A.4: Entering mathematical formulae

A3 contains a mathematical formula. ALL mathematical formulae MUST begin with "=" (the equal sign) so OpenOffice can tell that you are entering a mathematical formula rather than a number or text. =a1+a2 was entered into cell A3. Note that because cell A3 is selected, its contents are displayed in the "Input" line. The formula refers to cells A1 and A2, and looks to the current values in these cells when performing the function. Now any new values (numbers) entered into cells A1 and/or A2 will cause the formula in cell A3 to be executed.

ALL spreadsheets work this way!

ENTERING STATISTICAL FORMULAE

Now consider the illustration in Fig. A.5. In cells D2 through D11 are ten randomly measured customer opinions (cell D1 is a descriptive label). Let's say you want to use the OpenOffice statistical function COUNT to count the ten customer opinions and place the result in cell D12. (Note: this is a very simple example for illustrative purposes only.)

First, click on cell D12 to select it.

To start the process, first click on “Insert” to open the Insert menu. Next, point to “Function”, then click on it. This will start the function insertion process into the selected cell. You can see now why cell D12 was selected before the insertion process was started. (Actually, any cell not in the range of data cells would work.) The process initiation is illustrated in Fig. A.5.

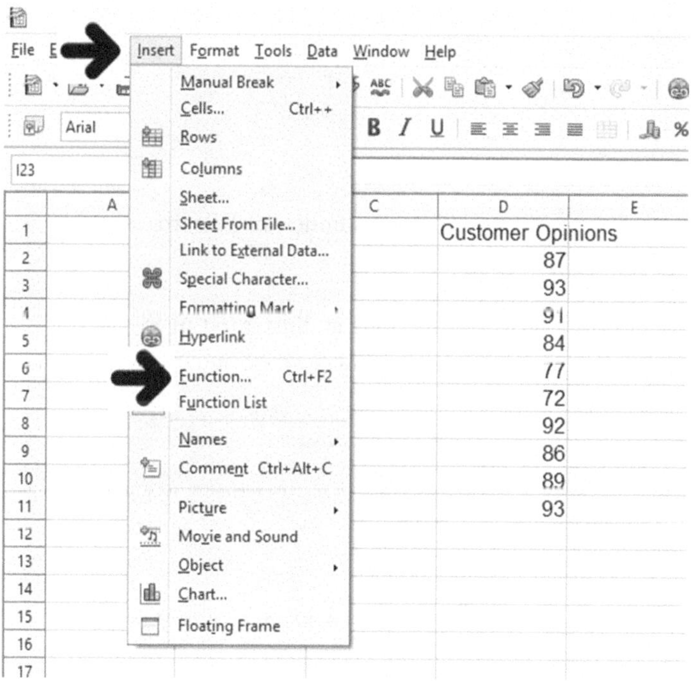

Figure A.5: Entering statistical formulae

The “Function Wizard” window, illustrated in Fig. A.6, will open when you click on “Function”. In the upper left of the “Function Wizard” window is word “Category”, with the word “All” highlighted in blue, and a down arrow to its right. Click on the down arrow to open a list of function categories. Fourth from bottom of that list is the “Statistics” category. Click on “Statistics” to open a window of only statistics functions.

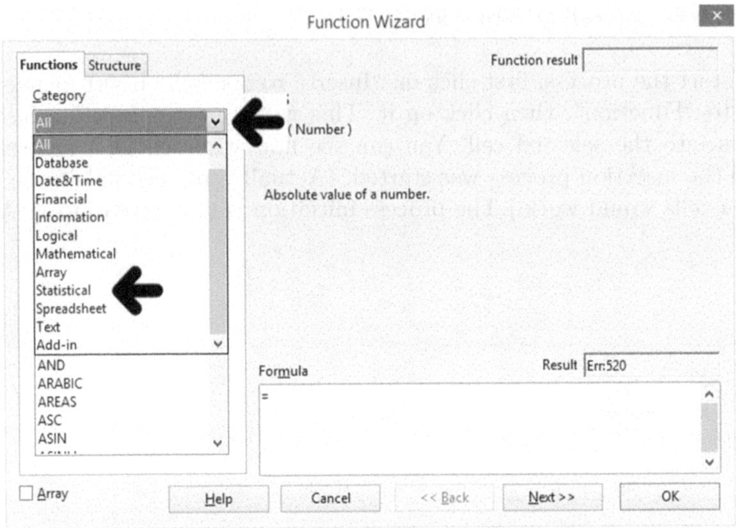

Figure A.6: The Function Wizard

The list of only statistics functions is illustrated in Fig. A.7.

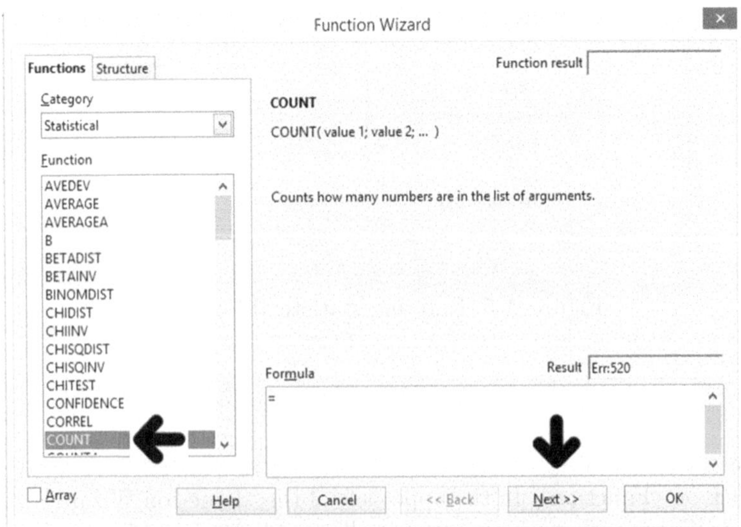

Figure A.7: Statistics functions

Notice that the "COUNT" function is near the bottom of this window. Click on the word "COUNT" to highlight and select it, then on "Next" in the lower right corner to start the COUNT function process. This will open the "Function Wizard" window. Note that the beginning of the COUNT function appears in the input line. The cursor will be in the "value 1" window. Type in the cell beginning and ending addresses of the cells you want to count. Enter the beginning cell address, a colon (:), then the ending cell address. Finally, click on "OK" to finalize the process (Fig. A.8).

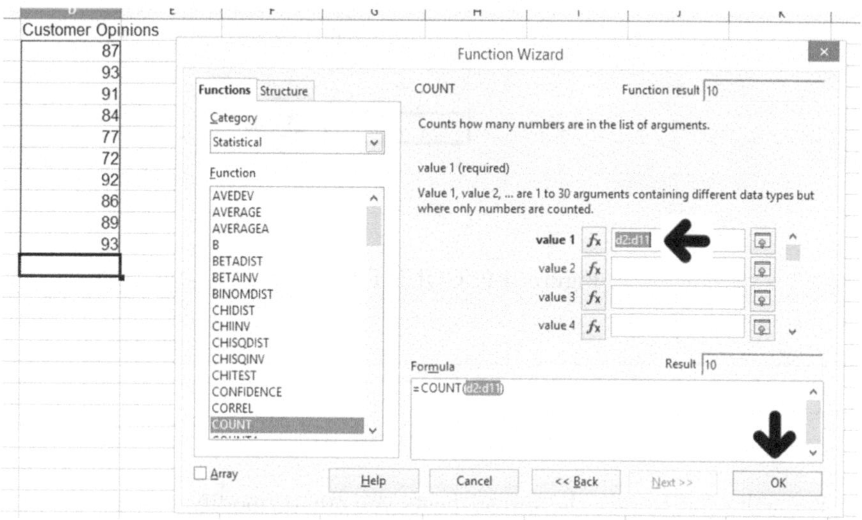

Figure A.8: Using the Function Wizard

ALL OpenOffice statistics functions are entered this way.

You can, of course, type the entire statistics function into the selected cell if you want to. Argument values are explained within the text. But this method is ultimately easier (Fig. A.9).

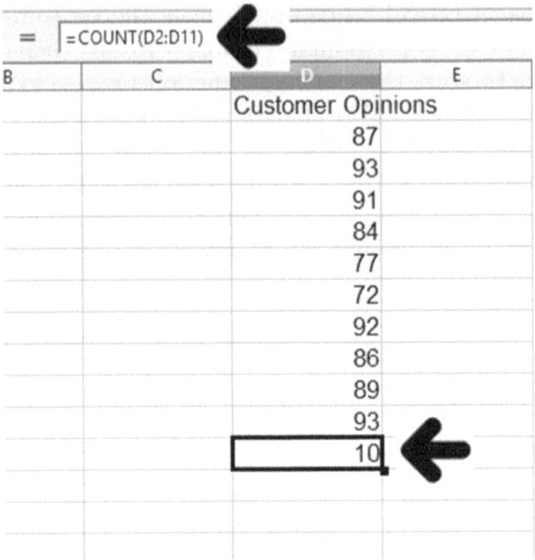

Figure A.9: COUNT function

SAVING YOUR WORK

You will want to save the spreadsheets you create/modify. Doing this with OpenOffice is easy.

First, it is STRONGLY recommended that you use Windows to create a unique directory or sub-directory where all spreadsheets will be saved. You will then know where to look for your spreadsheets. "openoffice spreadsheets" is the directory name used in this example.

Let's say that you have just created spreadsheet illustrated bellow and now want to save it. You can specify any spreadsheet name, but specifying a "meaningful" name will assist you when you want to recall it. Because the spreadsheet is about counting opinions, you name it "first count" because it is about counts and because it's the first spreadsheet about counts you created. Spreadsheet names can be any length and can be as descriptive as you desire.

Start the Save process by clicking on "Save" icon illustrated in Fig. A.10. The "select a File" window, illustrated in Fig. A.11, will open.

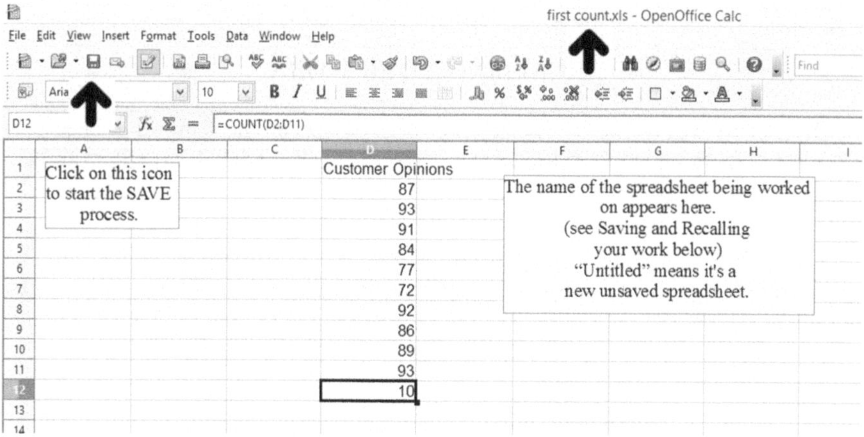

Figure A.10: Starting the SAVE process

Enter the name you want—"first count" in this example. Finally, click on "Save" to save the spreadsheet.

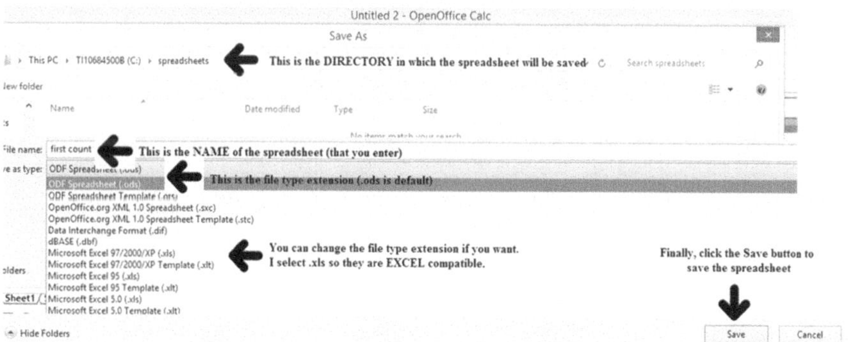

Figure A.11: Completing the SAVE process

RECALLING YOUR WORK

Often you will want to recall (or open) a spreadsheet upon which you were previously working. This is also easy to do. First, click on the "Open " icon to start the process (Fig. A.12).

The "Open" window, illustrated in Fig. A.13, will open. Click on the spreadsheet you want to open—it will be highlighted. Click on "Open" to open it.

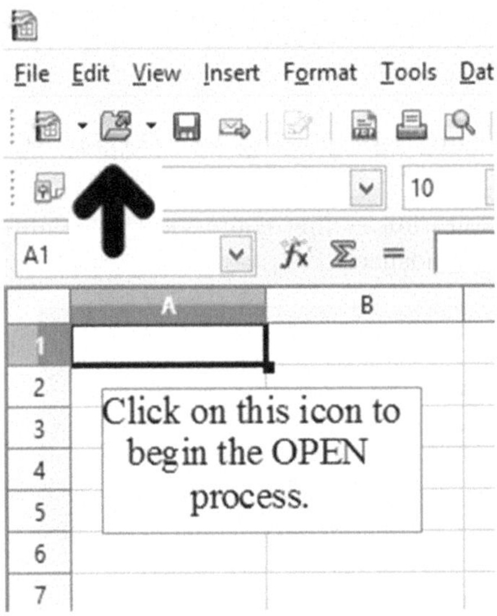

Figure A.12: Recalling your work

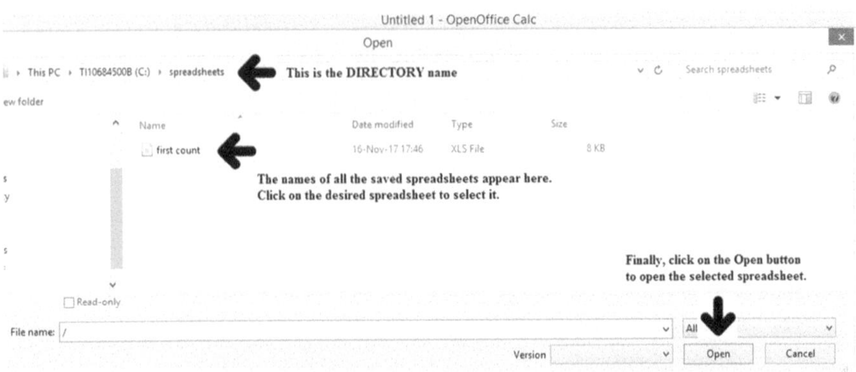

Figure A.13: Recalling your work—part 2

BACKUPS

You live and die by your backups. When, not if, your computer dies, you will need to have alternate access to your spreadsheets. You need to perform two types of backups regularly:

Internal backups—use Windows to copy all spreadsheets you have saved to some other directory. Being able to get to original, unchanged, uncorrupted spreadsheets will pay off.

External backups—use Windows to copy all spreadsheets you have saved to some external device, such as a USB flash drive. Being able to get to original, unchanged, uncorrupted spreadsheets will pay off.

Remember, if you perform backups, you cannot hurt anything, so play, play, play until you are comfortable!

The manufacturer's authorised representative in the EU is Springer Nature Customer Service Centre GmbH, Europaplatz 3, 69115 Heidelberg, Germany. If you have any concerns regarding our products, please contact ProductSafety@springernature.com

Printed and bound by CPI Group (UK) Ltd, Croydon, CR0 4YY
25/09/2026
02222857-0017